CLIMATE CHANGE AND ITS CAUSES, EFFECTS AND PREDICTION

CLIMATE CHANGE

LEGISLATIVE ISSUES AND ECONOMIC COSTS

CLIMATE CHANGE AND ITS CAUSES, EFFECTS AND PREDICTION

Climate Change and its Causes, Effects and Prediction

Climate Change

Legislative Issues and Economic Costs

Iyana Hendricks
Editor

snova
New York

NOTICE TO THE READER

Library of Congress Cataloging-in-Publication Data

ISBN: 978-1-53617-757-2

Published by Nova Science Publishers, Inc. † New York

CONTENTS

PREFACE

Chapter 1 will cover a brief history of U.S. climate change regulation; review the different types of regulation and legal actions that have been pursued in the national debate over GHGs; examine selected legal issues and next steps in related litigation; and address what these legal and regulatory developments mean for Congress.

The United States committed to providing financial assistance to developing countries for climate-change-related activities through the United Nations Framework Convention on Climate Change (UNFCCC) as reported in chapter 2.

The costs of recent weather disasters have illustrated the need for planning for climate change risks and investing in resilience.

Resilience is the ability to prepare and plan for, absorb, recover from, and more successfully adapt to adverse events, according to the National Academies of Science, Engineering, and Medicine. Chapter 3 reports that the federal government has not made measurable progress since 2017 to reduce fiscal exposure to climate change.

Chapter 4 focuses on the policy considerations and potential impacts of using a carbon tax or GHG emissions fee to control GHG emissions.

Administered by EPA, Superfund is the principal federal program for addressing sites containing hazardous substances. Chapter 5 reviews issues related to the impact of climate change on nonfederal NPL sites

A recent decision in the United States District Court for the District of Columbia Circuit has paused oil and gas exploration and production activity in certain leased areas of Wyoming and hinted at heightened requirements that the Bureau of Land Management (BLM) must satisfy to comply with the National Environmental Policy Act (NEPA) before issuing oil and gas leases. Specifically, the decision will require BLM to conduct a more thorough review of the potential climate change impacts of certain oil and gas leases before allowing the lessees to conduct drilling operations as discussed in chapter 6.

Chapter 7 summaries the content of the U.N. Framework Convention on Climate Change (UNFCCC) and its two subsidiary international treaties: the 1997 Kyoto Protocol (KP) and the 2015 Paris Agreement (PA).

The Green Climate Fund (GCF) was officially opened for capitalization at the U.N. Climate Summit in September 2014. Chapter 8 discusses how the funds were used.

Chapter 1 - On March 28, 2017, President Trump signed an executive order to encourage and promote energy development by modifying climate change policies. As the Trump Administration implements its environmental policies, various legal challenges to Obama Administration climate change regulations remain pending before courts. During the last term of the Obama Administration, the Environmental Protection Agency (EPA) and the National Highway and Traffic Safety Administration finalized a series of regulations to address emissions from cars, trucks, and their engines that may contribute to climate change. In addition, EPA finalized regulations pursuant to its authority under the Clean Air Act (CAA) to reduce GHG emissions from stationary sources such as power plants, GHG-emitting oil and gas sources, and landfills. Various stakeholders have challenged a majority of these rules generally contesting the scope of EPA's authority and its methods for regulating GHG emissions. In addition to the CAA, other environmental statutes such as the Endangered Species Act and the National Environmental Policy Act require federal agencies to consider climate change in their actions and decisions. The extent to which agencies may consider climate change effects and rely on predictive models, studies, and assumptions, however, has been challenged in court. Federal agencies are also required to consider the cost of GHG emissions in their rulemakings and environmental reviews. As the Trump Administration implements its policies on climate change, stakeholders may sue to ensure compliance with laws and judicial precedent that require consideration of climate change effects or costs. Climate change litigation may potentially increase as some stakeholders seek to reduce GHG emissions and address climate change effects. In the past, plaintiffs have had little success in using federal common law nuisance claims to force private entities to reduce their GHG emissions or pay damages for alleged injuries caused by their emissions. In 2011, the Supreme Court determined that these claims were displaced when Congress granted EPA authority to regulate GHG emissions under the CAA. If Congress amends the CAA to remove EPA's authority, plaintiffs may seek to reintroduce these common law nuisance claims. However, they will likely face jurisdictional barriers that may be difficult to overcome. Federal courts often do not reach the merits of climate change suits due to threshold procedural and jurisdictional barriers, such as whether a plaintiff or petitioner has the right to bring a lawsuit in the first place or whether the court has jurisdiction over a type of claim. These difficult procedural and jurisdictional barriers are at the center of a recent case claiming that the government has a duty to safeguard certain natural resources for the benefit of the public and that duty compels the government to address climate change. This chapter will cover a brief history of U.S. climate change regulation; review the different types of regulation and legal actions that have been pursued in the national debate over GHGs; examine selected legal issues and next steps in related litigation; and address what these legal and regulatory developments mean for Congress.

Chapter 2 - The United States committed to providing financial assistance to developing countries for climate-change- related activities through the United Nations Framework Convention on Climate Change (UNFCCC). The United States ratified the UNFCCC in 1992 with the advice and consent of the Senate (U.S. Treaty Number: 102-38). Among the obligations outlined in Article 4 of the UNFCCC, higher-income Parties (i.e., those listed in Annex II of the Convention, which were members of the Organization for Economic Cooperation and Development in 1992) sought to provide unspecified amounts of "financial resources, including for the transfer of technology, needed by developing countries to meet the agreed full incremental costs of implementing measures" to meet their general commitments under the UNFCCC. Further, "the implementation of these commitments shall take into account the need for adequacy and predictability in the flow of funds and the importance of appropriate burden sharing among the developed country Parties." Over the past several decades, and to varying degrees, the United States has delivered financial and technical assistance for climate-change-related activities in the developing world through a variety of bilateral and multilateral programs. U.S.-sponsored bilateral assistance has come through programs at the U.S. Agency for International Development (USAID); the Millennium Challenge Corporation; the Environmental Protection Agency; and the U.S. Departments of State, Agriculture, Commerce, and Energy, among others. U.S.-sponsored multilateral assistance has come through contributions by the U.S. Departments of State and the Treasury to environmental funds at various international financial institutions and organizations such as the Global Environment Facility, the Green Climate Fund, the U.N. Development Program, the U.N. Environment Program, the UNFCCC's Special Climate Change Fund, the UNFCCC's Least Developed Country Fund, the World Bank's Climate Investment Funds, and the World Bank's Forest Carbon Partnership Facility, among others. Each fund has its own mission and particular capacities.

Chapter 3 - Since 2005, federal funding for disaster assistance is at least $450 billion, including approximately $19.1 billion in supplemental appropriations signed into law on June 6, 2019. In 2018 alone, there were 14 separate billion-dollar weather and climate disaster events across the United States, with a total cost of at least $91 billion, according to the National Oceanic and Atmospheric Administration. The U.S. Global Change Research Program projects that disaster costs will likely increase as certain extreme weather events become more frequent and intense due to climate change. The costs of recent weather disasters have illustrated the need for planning for climate change risks and investing in resilience. Resilience is the ability to prepare and plan for, absorb, recover from, and more successfully adapt to adverse events, according to the National Academies of Science, Engineering, and Medicine. Investing in resilience can reduce the need for far more costly steps in the decades to come. Since February 2013, GAO has included *Limiting the Federal Government's Fiscal Exposure by Better Managing Climate Change Risks* on its list of federal program areas at high risk of vulnerabilities to fraud, waste, abuse, and mismanagement or most in need of transformation. GAO updates this list every 2 years. In March 2019, GAO reported that the federal government had not made measurable progress since 2017 to reduce fiscal exposure to climate change. This testimony—based on reports GAO issued from October 2009 to October 2019—discusses 1) what is known about the potential economic effects of climate change in the United States and the extent to which this information could help federal decision makers manage climate risks across the federal government, (2) the fiscal exposure facing the federal government due to climate risks and current efforts to address that exposure, (3) the extent to which the federal government has invested in resilience to climate change impacts, and (4) how the federal government could reduce fiscal exposure to the effects of climate change. GAO had made 62 recommendations related to the Limiting the Federal Government's Fiscal Exposure by Better Managing Climate Change Risks high-risk area. As of December 2018, 25 of those recommendations remained open.

Chapter 4 - The U.S. Fourth National Climate Assessment, released in 2018, concluded that "the impacts of global climate change are already being felt in the United States and are projected to intensify in the future—but the severity of future impacts will depend largely on actions taken to reduce greenhouse gas [GHG] emissions and to adapt to the changes that will occur." Members of Congress and stakeholders articulate a wide range of perspectives over what to do, if anything, about GHG emissions, future climate change, and related impacts. If Congress were to consider establishing a program to reduce GHG emissions, one option would be to attach a price to GHG emissions with a carbon tax or GHG emissions fee. In the 115th Congress, Members introduced nine bills to establish a carbon tax or emissions fee program. However, many Members have expressed their opposition to such an approach. In particular, in the 115th Congress, the House passed a resolution "expressing the sense of Congress that a carbon tax would be detrimental to the United States economy." Multiple economic studies have estimated the emission reductions that particular carbon tax would achieve. For example, a 2018 study analyzed various impacts of four carbon tax rate scenarios: a $25/metric ton of CO_2 and $50/metric ton of CO_2 carbon tax, increasing annually by 1% and 5%. The study concluded that each of the scenarios would likely achieve the U.S. GHG emission reduction target pledged under the international Paris Agreement (at least in terms of CO2 emissions). A carbon tax system would generate a new revenue stream, the magnitude of which would depend on the scope and rate of the tax, among other factors. In 2018, the Congressional Budget Office (CBO) estimated that a $25/metric ton carbon tax would yield approximately $100 billion in its first year. CBO projected that federal revenue would total $3.5 trillion in FY2019. Policymakers would face challenging decisions regarding the distribution of the new carbon tax revenues. Congress could apply revenues to support a range of policy objectives but would encounter trade-offs among the objectives. The central trade-offs involve minimizing economy-wide costs, lessening the costs borne by specific groups—particularly low-income households and displaced workers in certain industries (e.g., coal mining)—and supporting other policy objectives. A primary argument against a carbon tax regards it potential economy-wide impacts, often measured as impacts to the U.S. gross domestic product (GDP). Some may argue that projected impacts should be compared with the climate benefits achieved from the program as well as the estimated costs of taking no action. The potential impacts would depend on a number of factors, including the program's magnitude and design and, most importantly, the use of carbon tax revenues. In general, economic literature finds that some of the revenue applications would reduce the economy-wide costs from a carbon tax but may not eliminate them entirely. In addition, some studies cite particular economic modeling scenarios in which certain carbon tax revenue applications produce a net increase in GDP compared to a baseline scenario. These scenarios involve using carbon tax revenues to offset reductions in other tax rates (e.g., corporate income or payroll taxes). Although economic models generally indicate that these particular revenue applications would yield the greatest benefit to the economy overall, the models also find that lower-income households would likely face a disproportionate impact under such an approach. As lower-income households spend a greater proportion of their income on energy needs (electricity, gasoline), these households are expected to experience disproportionate impacts from a carbon tax if revenues were not recycled back to them in some fashion (e.g., lump-sum distribution). A price on GHG emissions could create a competitive disadvantage for some industries, particularly "emission-intensive, trade-exposed industries." Policymakers have several options to address this concern, including establishing a "border carbon adjustment" program, which would levy a fee on imports from countries without comparable GHG reduction programs. Alternatively, policymakers could allocate (indefinitely or for a period of time) some of the carbon tax revenues to selected industry sectors or businesses. Relatedly, a carbon tax system is projected to disproportionately impact fossil fuel industries, particularly coal, and the communities that rely on their employment. To alleviate these impacts, policymakers may consider using some of the revenue to provide transition assistance to employees or affected communities.

Chapter 5 - Administered by EPA, Superfund is the principal federal program for addressing sites containing hazardous substances. EPA lists some of the most seriously contaminated sites—most of which are nonfederal—on the NPL and has recorded over 500 contaminants, including arsenic and lead, at those sites. Climate change may make some natural disasters more frequent or more intense, which may damage NPL sites and potentially release contaminants, according to the Fourth National Climate Assessment. GAO was asked to review issues related to the impact of climate change on nonfederal NPL sites. This chapter examines, among other objectives, (1) what available federal data suggest about the number of nonfederal NPL sites that are located in areas that may be impacted by selected climate change effects and (2) the extent to which EPA has managed risks to human health and the environment from the potential impacts of climate change effects at such sites. GAO analyzed available federal data; reviewed laws, regulations, and documents; interviewed federal officials and stakeholders; visited three nonfederal NPL sites that experienced natural disasters; and compared EPA actions to manage risk to GAO's six essential elements of enterprise risk management.

Chapter 6 - A recent decision in the United States District Court for the District of Columbia Circuit has paused oil and gas exploration and production activity in certain leased areas of Wyoming and hinted at heightened requirements that the Bureau of Land Management (BLM) must satisfy to comply with the National Environmental Policy Act (NEPA) before issuing oil and gas leases. Specifically, the decision will require BLM to conduct a more thorough review of the potential climate change impacts of certain oil and gas leases before allowing the lessees to conduct drilling operations.

Chapter 7 - The United Nations Framework Convention on Climate Change (UNFCCC) has been the principle forum for cooperation among nations on greenhouse gas (GHG)-induced climate change since its adoption in 1992. Its objective is "to stabilize greenhouse gas concentrations in the atmosphere at a level that will prevent dangerous human interference with the climate system, in a time frame which allows ecosystems to adapt naturally and enables sustainable development." Stabilizing GHG concentrations in the atmosphere requires that the balance of "gross" emissions of GHG minus the removals of GHG from the atmosphere reach "net zero." Two principles agreed in the UNFCCC are that (1) Parties should act "on the basis of equity and in accordance with their common but differentiated responsibilities and respective capabilities" and (2) developed country Parties should take the lead in combating climate change. The bifurcation of responsibilities among Parties into developed (Annex I) and developing countries has been a major point of contention. Annex I Parties, including the United States, had stronger obligations, such as more rigorous reporting and reviews. A subset listed in Annex II, including the United States, committed to provide agreed financial resources and technology transfers. The commitments are qualitative and collective, not binding on individual Parties. The first subsidiary agreement to the UNFCCC was the 1997 Kyoto Protocol (KP), which entered into force in 2005. The United States signed but did not ratify the KP and so is not a Party. The developed Parties agreed to reduce GHG emissions by 5% below their 1990 levels, with different targets for each Party. In 2009, a political declaration, the Copenhagen Accord, led to explicit pledges from many Parties to mitigate GHG, though they remained bifurcated as Annex I and non-Annex I (i.e., developing countries) by both the type of action and the frequency and format of the reporting requirements. In 2010, the Cancun agreements took note of a Copenhagen pledge by developed country Parties to jointly mobilize $100 billion per year by 2020. Funds provided "may come from a wide variety of sources, public and private, bilateral and multilateral, including alternative sources." The Paris Agreement (PA) is the second major subsidiary agreement under the UNFCCC. The PA defines a collective, long- term objective to hold the GHG-induced increase in temperature to well below 2o Celsius (C) and to pursue efforts to limit the temperature increase to 1.5o C above the pre-industrial level. In the PA, for the first time under the UNFCCC, all Parties participate in a common framework with common guidance, though some Parties are allowed limited flexibility. The negotiators intended the PA to be legally binding on its Parties, though not all provisions are mandatory. All Parties must submit "Nationally Determined Contributions" (NDCs) containing nonbinding pledges to mitigate GHG emissions. The Parties are to update or submit new NDCs by 2020 and every five years thereafter. Each successive NDC of a Party "will represent a progression" and "reflect its highest possible ambition, reflecting its common but differentiated responsibilities and respective capabilities, in light of different national circumstances." The PA reiterates the obligation in the UNFCCC for developed country Parties to seek to mobilize financial support to assist developing country Parties with climate change mitigation and adaptation efforts, encouraging all Parties to provide financial support voluntarily. The decision to carry out the PA calls for continuing the Cancun collective mobilization through 2025. The Parties agree to set, prior to their 2025 meeting, a new collective, quantified goal of not less than $100 billion annually to assist developing country Parties. President Trump announced his intention in 2017 to withdraw the United States from the PA as soon as it was eligible. The U.S. Department of State notified the United Nations of U.S. withdrawal on November 4, 2019. The withdrawal takes effect on November 4, 2020, unless the U.S. government postpones or rescinds the withdrawal. A Party may reenter the PA 30 days after depositing notice that it has ratified, accepted, or acceded to the PA.

Chapter 8 - The Green Climate Fund (GCF) was officially opened for capitalization at the U.N. Climate Summit in September 2014. (See CRS In Focus IF10382, *The Green Climate Fund (GCF)* for a history and overview of the fund.) The GCF's initial resource mobilization (IRM) lasted from 2015 to 2018. As of the most recent reporting, April 30, 2019, the GCF had raised over $10.2 billion (all amounts are in U.S. $ equivalent) in signed pledges from 48 countries/regions/cities during the IRM period.

In: Climate Change
Editor: Iyana Hendricks

ISBN: 978-1-53617-757-2

Chapter 1

U.S. CLIMATE CHANGE REGULATION AND LITIGATION: SELECTED LEGAL ISSUES*

Linda Tsang

ABSTRACT

On March 28, 2017, President Trump signed an executive order to encourage and promote energy development by modifying climate change policies. As the Trump Administration implements its environmental policies, various legal challenges to Obama Administration climate change regulations remain pending before courts. During the last term of the Obama Administration, the Environmental Protection Agency (EPA) and the National Highway and Traffic Safety Administration finalized a series of regulations to address emissions from cars, trucks, and their engines that may contribute to climate change. In addition, EPA finalized regulations pursuant to its authority under the Clean Air Act (CAA) to reduce GHG emissions from stationary sources such as power plants, GHG-emitting oil and gas sources, and landfills.

---

* This is an edited, reformatted and augmented version of Congressional Research Service. Publication No. R44807, dated April 13, 2017.

Various stakeholders have challenged a majority of these rules generally contesting the scope of EPA's authority and its methods for regulating GHG emissions.

In addition to the CAA, other environmental statutes such as the Endangered Species Act and the National Environmental Policy Act require federal agencies to consider climate change in their actions and decisions. The extent to which agencies may consider climate change effects and rely on predictive models, studies, and assumptions, however, has been challenged in court. Federal agencies are also required to consider the cost of GHG emissions in their rulemakings and environmental reviews. As the Trump Administration implements its policies on climate change, stakeholders may sue to ensure compliance with laws and judicial precedent that require consideration of climate change effects or costs.

Climate change litigation may potentially increase as some stakeholders seek to reduce GHG emissions and address climate change effects. In the past, plaintiffs have had little success in using federal common law nuisance claims to force private entities to reduce their GHG emissions or pay damages for alleged injuries caused by their emissions. In 2011, the Supreme Court determined that these claims were displaced when Congress granted EPA authority to regulate GHG emissions under the CAA. If Congress amends the CAA to remove EPA's authority, plaintiffs may seek to reintroduce these common law nuisance claims. However, they will likely face jurisdictional barriers that may be difficult to overcome.

Federal courts often do not reach the merits of climate change suits due to threshold procedural and jurisdictional barriers, such as whether a plaintiff or petitioner has the right to bring a lawsuit in the first place or whether the court has jurisdiction over a type of claim. These difficult procedural and jurisdictional barriers are at the center of a recent case claiming that the government has a duty to safeguard certain natural resources for the benefit of the public and that duty compels the government to address climate change.

This chapter will cover a brief history of U.S. climate change regulation; review the different types of regulation and legal actions that have been pursued in the national debate over GHGs; examine selected legal issues and next steps in related litigation; and address what these legal and regulatory developments mean for Congress.

Introduction

In 2013, President Obama released a Climate Action Plan (CAP) to reduce greenhouse gases (GHG) emissions and to encourage adaptation to expected climate change.[1] Subsequently, the Obama Administration implemented the CAP pursuant to presidential powers and existing statutory authority. The President, the U.S. Environmental Protection Agency (EPA), and other agencies issued various executive orders, guidance, and regulations to reduce GHG emissions from motor vehicles,[2] new and existing power plants,[3] facilities in the oil-and-gas sector,[4] and other GHG-emitting sources and federal facilities.[5]

On March 28, 2017, President Trump signed Executive Order 13783, which rescinds the CAP and requires agencies to review their existing regulations and "appropriately suspend, revise, or rescind those that unduly burden" domestic energy production and use.[6] The executive order, entitled *Promoting Energy Independence and Economic Growth*, rescinds or withdraws various Obama Administration orders and directions that were intended to address climate change and regulate GHG emissions, including the following:[7]

- Presidential Orders/Reports
  - Executive Order 13653, Preparing the United States for the Impacts of Climate Change (Nov. 1, 2013)

---

[1] Exec. Off. of the President, The President's Climate Action Plan (2013), *available at* https://obamawhitehouse.archives.gov/sites/default/files/image/president27sclimateactionplan.pdf. For additional background on the Climate Action Plan, see CRS Report R43120, *President Obama's Climate Action Plan*, coordinated by Jane A. Leggett.

[2] *See infra* "Mobile Sources under the Clean Air Act" section.

[3] *See infra* "Power Plants" section.

[4] *See infra* "Regulation of Methane Emissions from Stationary Sources" section.

[5] *See, e.g.*, Exec. Order No. 13693, 80 Fed. Reg. 15,871 (Mar. 25, 2015).

[6] *See* Exec. Order No. 13783, 82 Fed. Reg. 16, 093 (Mar. 31, 2017) (signed on Mar. 28, 2017),https://www.federalregister.gov/d/2017-06576. Michael Phillis, Trump Orders EPA To Undo Clean Power Plan, LAW360, Mar. 28, 2017, https://www.law360.com/energy/articles/906980/breaking-trump-orders-epa-to-undo-cleanpower-plan.

[7] *Id.*

  - Power Sector Carbon Pollution Standards, Pres. Memo. (June 25, 2013)
  - Mitigating Impacts on Natural Resources from Development and Encouraging Related Private Investment, Pres. Memo. (Nov. 3, 2015)
  - Climate Change and National Security, Pres. Memo. (Sept. 21, 2016)
  - The President's Climate Action Plan (June, 2013)
  - Climate Action Plan Strategy to Reduce Methane Emissions (March 2014)
- Agency Guidance
  - Final Guidance for Federal Departments and Agencies on Consideration of Greenhouse Gas Emissions and the Effects of Climate Change in National Environmental Policy Act Reviews, 81 Fed. Reg. 51,866 (Aug. 5, 2016)
  - All technical documents related to the Social Cost of Carbon
- Agency Orders
  - All moratoria on federal land coal leasing activities related to Order 3338, Discretionary Programmatic Environmental Impact Statement (PEIS) to Modernize the Federal Coal Program (Jan. 15, 2016)

The executive order also directs EPA to review GHG regulations for power plants and methane regulations for existing oil and gas sources, and "as soon as practicable, suspend, revise, or rescind the guidance, or publish for notice and comment proposed rules suspending, revising, or rescinding those rules."[8]

Because the evolution and scope of U.S. climate change regulation and litigation are extensive, this chapter will focus on existing federal laws and regulations and will not address international agreements or legislation

---

[8] *Id.* §§4, 7.

related to climate change.[9] This chapter will provide a brief history of U.S. climate change regulation; review the different types of regulation and legal actions that have been pursued in the national debate over GHGs; examine selected legal issues and related litigation; and address what these legal and regulatory developments mean for Congress.

BACKGROUND

History of U.S. Climate Change Regulation and Litigation

Since the late 1970s, Congress has provided authority and funds for federal agencies to research and better understand climate science. In 1978, Congress passed the National Climate Program Act (NCPA) that created the National Climate Program in the Department of Commerce.[10] The NCPA required the program it established to assess climate change effects, improve understanding of climate change science, and collect global and domestic climate data.[11] Congress subsequently passed additional legislation authorizing federal climate change research, interagency collaboration, and collection of GHG emissions data.[12] These statutes authorized federal agencies to investigate climate change and its potential impacts, but did not direct agencies to reduce GHG emissions.

---

[9] For information related to international climate change agreements, see CRS Report R44609, *Climate Change: Frequently Asked Questions about the 2015 Paris Agreement*, by Jane A. Leggett and Richard K. Lattanzio, and CRS Report R44761, *Withdrawal from International Agreements: Legal Framework, the Paris Agreement, and the Iran Nuclear Agreement*, by Stephen P. Mulligan.

[10] National Climate Program Act of 1978, P.L. 95-367, 92 Stat. 601 (1978) (as amended at 15 U.S.C. §§29.1-2908); 15 U.S.C. §2904(c).

[11] 15 U.S.C. §2904(d).

[12] *See, e.g.*, Global Change Research Act of 1990, 15 U.S.C. §§2921-2938 (establishing the Global Change Research Program to coordinate federal climate research to aid in "understand[ing] ... human-induced and natural processes of climate change"); Global Climate Change Prevention Act of 1990, 7 U.S.C. §6701 (directing the Dept. of Agriculture to study the effects of climate change on forestry and agriculture); Energy Policy Act of 1992, 42 U.S.C. §§13381 13388 (requiring the Secretary of Energy to prepare an annual inventory of GHG emissions).

Various executive orders and environmental, energy, and natural resources statutes address climate change in different ways. For example, under the Clean Air Act (CAA), EPA has sought to reduce GHG emissions from stationary and mobile sources.[13] Similarly, the Department of Energy has established efficiency standards for certain consumer products to reduce energy consumption,[14] while the Council on Environmental Quality has issued guidance on how agencies should consider potential GHG impacts resulting from proposed federal actions under the National Environmental Policy Act.[15]

Federal authority to regulate GHGs under existing law and policies has been challenged in court. Many of these challenges have focused on EPA's authority to regulate GHGs under the CAA. In 1998, EPA General Counsel Jonathan Cannon concluded in a memorandum that GHGs were air pollutants within the CAA definition of the term, and therefore could be regulated under the CAA.[16] On October 20, 1999, relying on the Cannon memorandum as well as on the CAA, a group of 19 organizations petitioned EPA to regulate GHG emissions from new motor vehicles under CAA Section 202,[17] which directs EPA to develop emission standards for "any air pollutant" from new motor vehicles "which, in his judgment cause[s], or contribute[s] to air pollution which may reasonably be anticipated to endanger public health or welfare."[18]

On August 28, 2003, EPA denied the petition[19] based on a new General Counsel memorandum that concluded that the CAA does not grant

---

[13] *See infra* "Stationary Sources under Clean Air Act" and "Mobile Sources under the Clean Air Act" sections.

[14] *See infra* "Social Cost of Carbon" section.

[15] *See infra* "National Environmental Policy Act" section.

[16] Memorandum from Jonathan Z. Cannon, EPA Gen. Counsel, on EPA's Authority to Regulate Pollutants Emitted by Electric Power Generation Sources, to Carol M. Browner, Admin., EPA (Apr. 10, 1998), http://www.law.umaryland.edu/environment/casebook/documents/epaco2memo1.pdf.

[17] The lead petitioner was the International Center for Technology Assessment (ICTA). The petition may be found at http://www.ciel.org/Publications/greenhouse_petition_EPA.pdf.

[18] *Id.*

[19] Control of Emissions From New Highway Vehicles and Engines, 68 Fed. Reg. 52,922 (Sept. 8, 2003). The agency argued that it lacked statutory authority to regulate GHGs: Congress "was well aware of the global climate change issue" when it last comprehensively amended the CAA in 1990, according to the agency, but "it declined to adopt a proposed amendment establishing binding emissions limitations." Massachusetts v. EPA, 549 U.S. 497 (2007).

EPA authority to regulate carbon dioxide (CO2) and other GHG emissions based on their climate change impacts.[20] Massachusetts, 11 other states, and various other petitioners challenged EPA's denial of the petition in a case that ultimately reached the Supreme Court.[21]

Massachusetts v. EPA and Its Effects

In 2007, the Supreme Court defined the contours of EPA's authority to regulate GHGs under the Clean Air Act in *Massachusetts v. EPA*.[22] In its 5-4 decision, the Supreme Court held that EPA has authority to regulate GHGs as "air pollutants" under the CAA.[23] The Court determined that the CAA authorized EPA to regulate GHG emissions from new motor vehicles because CO_2 and other GHGs "fit well within the CAA's capacious [and unambiguous] definition of 'air pollutant.'"[24] Rejecting EPA's argument that the agency could not regulate vehicle GHG emissions because to do so would conflict with the Department of Transportation's (DOT's) fuel economy standards, the Court explained that EPA's statutory obligation to protect public health and welfare under the CAA is "wholly independent of DOT's mandate to promote energy efficiency."[25] In addition, the Court held that CAA Section 202 does not allow EPA to refuse to regulate GHG emissions from new motor vehicles based on policy considerations, but that EPA must regulate GHG emissions if EPA finds that GHGs "cause or

---

[20] Memorandum from Robert E. Fabricant, Gen. Counsel, EPA, on EPA's Authority to Impose Mandatory Controls to Address Global Climate Change Under the Clean Air Act, to Marianne L. Horinko, Acting Admin., EPA (Aug. 28, 2003), http://bit.ly/2nMx5Dy.

[21] The D.C. Circuit, in a split decision, rejected the suit. *See* Massachusetts v. EPA, 415 F.3d 50, 56, 59-60 (2005) (Randolph, J., dissenting) (holding that EPA reasonably denied the petition based on scientific uncertainty and policy considerations).

[22] Massachusetts v. EPA, 549 U.S. 497 (2007). For additional discussion of this case, see CRS Report RS22665, *The Supreme Court's First Climate Change Decision: Massachusetts v. EPA*, by Alexandra M. Wyatt. Issues related to standing addressed by the Court are discussed in the "Standing Doctrine" section.

[23] *Id.*

[24] *Massachusetts v. EPA*, 549 U.S. at 529, 532. The majority held that "The Clean Air Act's sweeping definition of 'air pollutant' includes '*any* air pollution agent or combination of such agents, including *any* physical, chemical ... substance or matter which is emitted into or otherwise enters the ambient air.... ' ... Carbon dioxide, methane, nitrous oxide, and hydrofluorocarbons are without a doubt 'physical [and] chemical ... substances[s] which [are] emitted into ... the ambient air.' The statute is unambiguous." *Id.* at 528-29.

[25] *Id.*

contribute to air pollution that may reasonably be anticipated to endanger public health or welfare."[26]

Based on this reasoning, the Court identified three ways EPA could respond to the rulemaking petition: (1) make a finding that motor vehicle GHG emissions may "endanger public health or welfare" and issue emissions standards; (2) make a finding that there is no endangerment from these emissions; (3) or provide a "reasonable explanation" as to why the Agency cannot or will not make an endangerment finding.[27] Because EPA "offered no reasoned explanation for its refusal to decide whether greenhouse gases cause or contribute to climate change," the Court held that EPA's denial of the petition for rulemaking was arbitrary and capricious, reversed the D.C. Circuit opinion, and remanded the case to that court for further proceedings.[28]

Following the Supreme Court's decision in *Massachusetts v. EPA,* EPA found that GHG emissions from motor vehicles endanger public health and welfare[29] and subsequently issued a series of regulations under the CAA to reduce GHG emissions from both mobile and stationary sources. EPA has consequently regulated GHG emissions from new motor vehicles under CAA Section 202.[30] EPA has also regulated GHG emissions from various stationary sources, including new and existing fossil fuel-fired power plants and GHG-emitting sources from municipal landfills, and the oil and gas sector under other CAA sections.[31]

Most of EPA's GHG regulatory actions have been challenged in court. These challenges generally concern the scope of EPA's authority and its methods for regulating GHG emissions—not whether EPA has authority to

---

[26] *Id.* at 532-33.

[27] *Id.* at 533.

[28] *Id.* at 534-35. A four-justice dissent by Chief Justice Roberts in *Massachusetts v. EPA* disputed the majority's holding of standing. *Id.* at 535-49 (Roberts, C.J., dissenting). Justice Scalia, for the same four justices, issued a dissent, arguing that EPA may appropriately consider agency policy preferences in determining whether to issue or defer an endangerment finding. *Id.* at 535-49 (Scalia, J., dissenting). Justice Scalia also disputed the majority's holding that "air pollutant" in Clean Air Act §202 includes GHGs. *Id.* at 559-60. For further discussion of the standing issues in *Massachusetts v. EPA*, see "Standing Doctrine" section.

[29] *See infra* "Reducing GHG Emissions" section.

[30] *See infra* "Mobile Sources under the Clean Air Act" section.

[31] *See infra* "Stationary Sources under Clean Air Act" section.

regulate GHG emissions under the CAA, which the Supreme Court held in *Massachusetts v. EPA*. Although the Supreme Court has clarified EPA's authority to regulate GHGs under specific CAA programs in subsequent cases,[32] *Massachusetts v. EPA* remains the governing law.

TYPES OF GHG REGULATION AND RELATED LITIGATION

This chapter will focus on two types of GHG regulations: (1) laws that seek to reduce GHG emissions from stationary and mobile sources, and (2) laws that require that climate change effects be considered when the federal government takes certain actions. This section will review examples of these types of regulations and related legal challenges.

Reducing GHG Emissions

The majority of regulations addressing GHG emissions from stationary and mobile sources have been promulgated under the CAA. Following the *Massachusetts* decision, EPA began to explore "whether and how greenhouse gases could be effectively controlled under the Clean Air Act" in an advanced notice of proposed rulemaking (ANPR).[33] In 2009, EPA made two findings under CAA Section 202: (1) that GHGs currently in the atmosphere potentially endanger public health and welfare, and (2) that new motor vehicle emissions cause or contribute to that pollution (collectively these findings are known as the "endangerment finding").[34]

---

[32] *See, e.g.*, Util. Air Regulatory Group v. EPA,—U.S.—, 134 S. Ct. 2427, 2439-46 (2014); Am. Elec. Power Co. v. Connecticut, 564 U.S. 410 (2011). These cases are discussed in "Regulation of GHG Emissions from Stationary Sources through Permits" and "Common Law Nuisance and Political Question Doctrine" sections.

[33] Regulating Greenhouse Gas Emissions Under the Clean Air Act; Advance Notice of Proposed Rulemaking, 73 Fed. Reg. 44,354, 44,355 (July 30, 2008).

[34] 74 Fed. Reg. 66,496 (Dec. 15, 2009).

The endangerment finding triggered EPA's duty under CAA Section 202(a) to promulgate emission standards for new motor vehicles.[35]

Mobile Sources under the Clean Air Act

In general, the CAA divides mobile sources into three broad categories: on-road vehicles (cars, trucks, and buses); aircraft; and nonroad vehicles and engines. Following the *Massachusetts v. EPA* decision, EPA has exercised its authority under Title II of the CAA to address GHG emissions from some mobile sources. This section will focus on EPA's regulation of motor vehicles, and briefly discuss EPA's ability to regulate GHG emissions from aircrafts and other nonroad mobile sources and engines.[36]

Light-Duty Motor Vehicles

While EPA's endangerment finding was pending, President Obama announced that EPA would develop its GHG emission standards in coordination with the National Highway Traffic Safety Administration's (NHTSA) development of corporate average fuel economy (CAFE) standards, which NHTSA administers for new cars and light trucks.[37] By developing these standards jointly, the Obama Administration sought to create a national program to reduce GHG levels to levels similar to those the California Air Resources Board (CARB) had adopted earlier.[38] CARB further agreed to harmonize its motor vehicle GHG emissions standards with EPA's standards.[39]

---

[35] *Id.*

[36] For additional information regarding GHG regulations of motor vehicles under the Clean Air Act, see CRS Report R40506, *Cars, Trucks, Aircraft, and EPA Climate Regulations*, by James E. McCarthy and Richard K. Lattanzio.

[37] Press Release, White House, Off. of the Secretary, President Obama Announces National Fuel Efficiency Policy (May 19, 2009), https://obamawhitehouse.archives.gov/the-press-office/president-obama-announces-national-fuelefficiency-policy.

[38] Proposed Rulemaking To Establish Light-Duty Vehicle Greenhouse Gas Emission Standards and Corporate Average Fuel Economy Standards, 74 Fed. Reg. 49,468 (Sept. 28, 2009).

[39] Letter from Mary D. Nichols, Chairman, Ca. Air Res. Bd, to Lisa P. Jackson, Admin., EPA & Ray LaHood, Secretary, Dept. of Transp. (May 18, 2009), https://www.epa.gov/

EPA and NHTSA subsequently promulgated GHG emission and fuel economy standards in two phases for specific model years of light-duty motor vehicles. On May 7, 2010, EPA and NHTSA issued Phase 1 standards, which applied to Model Year (MY) 2012-2016 light-duty vehicles (cars, sport-utility vehicles, crossovers, minivans, and most pickups), setting a combined average emissions standard of 250 grams/mile of CO2 and a combined average fuel economy level of 34.1 mpg in MY2016.[40] On October 15, 2012, EPA and NHTSA issued Phase 2 standards, which applied to MY2017-2025 light-duty vehicles, to reduce further the combined average emissions to 163 grams/mile of CO2 by MY2025 and combined average fuel economy to a range of 40.3 to 41 mpg by MY2021.[41]

To allow EPA and NHTSA to modify the GHG standards, the Phase 2 MY2017-2025 light-duty vehicle regulations provided for a Midterm Evaluation (MTE) of the MY2022-2025 standards.[42] Under the regulations, EPA had to determine by April 1, 2018, whether the GHG standards for MY2022-2025 would still be appropriate given the latest available data and information, and could strengthen, weaken, or retain the promulgated standards.[43] On January 12, 2017, EPA issued a final decision finding that the standards remained appropriate and should be maintained for the model years.[44] EPA based its findings on a technical support document; a

---

sites/production/files/2016-10/documents/airresources-board.pdf. EPA officially granted California's waiver for its 2009 GHG standards on July 8, 2009. California State Motor Vehicle Pollution Control Standards; Notice of Decision Granting a Waiver of Clean Air Act Preemption for California's 2009 and Subsequent Model Year Greenhouse Gas Emission Standards for New Motor Vehicles, 74 Fed. Reg, 32,744 (July 8, 2009). For discussion of state preemption waivers under CAA Section 209, see "California Preemption Waiver."

[40] Light-Duty Vehicle Greenhouse Gas Emission Standards and Corporate Average Fuel Economy Standards; Final Rule, 75 Fed. Reg. 25,323, 25,329-30 (May 7, 2010).

[41] 2017 and Later Model Year Light-Duty Vehicle Greenhouse Gas Emissions and Corporate Average Fuel Economy Standards; Final Rule, 77 Fed. Reg. 62,624, 62,638-39 (Oct. 15, 2012).

[42] 40 C.F.R. §86.1818-12(h). For more information on the MTE and determination, see CRS Insight IN10619, *EPA's Mid-Term Evaluation of Vehicle Greenhouse Gas Emissions Standards*, by Richard K. Lattanzio.

[43] *Id.*

[44] EPA, FINAL DETERMINATION ON THE APPROPRIATENESS OF THE MODEL YEAR 2022-2025 LIGHT-DUTY VEHICLE GREENHOUSE GAS EMISSIONS STANDARDS

previously released Draft Technical Assessment Report (which was issued jointly by EPA, NHTSA, and CARB); and input from the auto industry and other stakeholders.[45]

EPA has been criticized for finalizing the MTE determination prior to the April 1, 2018, deadline and separate from NHTSA's rulemaking on CAFE standards for MY2022-2025.[46] On March 13, 2017, the Alliance of Automobile Manufacturers filed a petition for review of EPA's final MTE determination in the D.C. Circuit, claiming that EPA's determination is "in excess of EPA's statutory authority, contrary to the CAA, arbitrary and capricious, and otherwise contrary to law."[47] Supporters of EPA's MTE determination have urged EPA not to reopen or withdraw the determination, arguing that the GHG standards are "economically feasible and technological achievable."[48] On March 24, 2017, CARB voted unanimously to retain California's vehicle GHG standards for MY2022-2025 for light-duty vehicles based on its own mid-term evaluation.[49]

On March 15, 2017, EPA and DOT announced that EPA would reconsider its MTE determination[50] and would coordinate with NHTSA in order to harmonize NHTSA's CAFE standards with EPA's GHG emission

---

UNDER THE MIDTERM EVALUATION (2017), https://www.epa.gov/sites/ production/files/2017-01/documents/420r17001.pdf.

[45] *Id.* at 2. The draft and final determinations and supporting documents can be found on EPA's website at https://www.epa.gov/regulations-emissions-vehicles-and-engines/midterm-evaluation-light-duty-vehicle-greenhousegas-ghg.

[46] *Id.* at 27. Some industry groups have asked the Trump Administration to revisit EPA's determination—citing a "rush" to finalize a determination that contradicted the objectives of a unified national program GHG and CAFE standards. *See, e.g.*, Letter from Alliance of Auto. Mfrs. et. al., to Mike Pence, Vice-President Elect (Dec. 22, 2016), https://www.uschamber.com/letter/multi-association-letter-trump-transition-team-proposed-determination-midtermevaluation-ghg; Camille von Kaenel, *Automakers ask Pruitt to Revisit Obama's Emissions Rules*, E&E NEWS, Feb. 22, 2017, http://www.eenews.net/climatewire/stories/1060050394/search?keyword=alliance+automobile+manufacturers.

[47] Petition for Review, Alliance of Auto. Mfrs. v. EPA, No. 17-1086 (D.C. Cir. Mar. 13, 2017).

[48] *See, e.g.*, Letter from Edward J. Markey, at. al, Senator, to Scott Pruitt, Admin., EPA (Mar. 7, 2017), http://www.markey.senate.gov/imo/media/doc/EPA%20auto%20emissions%20final%20determination%20letter.pdf.

[49] *California Air Regulators Vote to Keep Tough Fuel Standards*, ASSOCIATED PRESS, Mar. 24, 2017, http://hosted2.ap.org/APDEFAULT/b2f0ca3a594644ee9e50a8ec4ce2d6de/Article_2017-03-24-US-California-FuelStandards/id-dce9c6442699433ebde1497672537f30.

[50] *See* Notice of Intention To Reconsider the Final Determination of the Mid-Term Evaluation of Greenhouse Gas Emissions Standards for Model Year 2022–2025 Light Duty Vehicles, 82 Fed. Reg. 14,671 (Mar. 22, 2017) (signed on Mar. 3, 2017).

standards.[51] EPA intends to issue a new MTE determination by April 1, 2018.[52] Given that EPA is reconsidering its MTE determination, the Alliance of Automobile Manufacturers has moved to dismiss its lawsuit.[53]

Medium and Heavy-Duty Trucks

Section 202(a) of the CAA requires EPA to set "standards applicable to the emission of any air pollutant from *any class or classes of new motor vehicles or new motor vehicle engines*, which in the agency's judgment cause, or contribute to, air pollution which may reasonably be anticipated to endanger public health or welfare."[54] In the endangerment finding, EPA specifically identified medium- and heavy-duty trucks as contributing to endangerment caused by GHG emissions.[55] Similar to the light-duty vehicles standards, EPA and NHTSA[56] issued GHG emissions and fuel economy standards for medium- and heavy-duty trucks in two phases. On September 15, 2011, EPA and NHTSA issued Phase 1 standards for MY2014-2018 medium- and heavy-duty trucks.[57] On October 25, 2016, EPA and NHTSA issued Phase 2 standards for MY2021-2027 medium- and heavy-duty trucks and engines and MY2018-MY2027 trailers.[58]

---

[51] *See id.* at 4 (citing 40 C.F.R. §86.1818-12(h)(1)(vii), which states that EPA "shall consider the information available on the factors relevant to setting greenhouse gas emission standards ... including but not limited to: ... [t]he impact of the greenhouse gas emission standards on the Corporate Average Fuel Economy standards and a national harmonized program").

[52] *Id.*

[53] Motion of Petitioner to Dismiss Petition for Review, Alliance of Auto. Mfrs. v. EPA, No. 17-1086, No. 17-1086 (D.C. Cir. Mar. 20, 2017).

[54] 42 U.S.C. §7521(a)(1) (emphasis added).

[55] Endangerment and Cause or Contribute Findings for Greenhouse Gases Under Section 202(a) of the Clean Air Act, 74 Fed. Reg. 66,496, 66,537 (Dec. 15, 2009).

[56] Separate from EPA's authority to set GHG standards, the Energy Independence and Security Act of 2007 directed NHTSA to study the potential for fuel efficiency standards for medium- and heavy-duty trucks, and, if feasible, set efficiency standards reflecting the "maximum feasible improvement." 49 U.S.C. §32902(k)(2). Similar to light-duty vehicles standards, in order to harmonize their GHG and fuel efficiency requirements, EPA and NHTSA have cooperated on the setting of motor vehicle standards.

[57] Greenhouse Gas Emissions Standards and Fuel Efficiency Standards for Medium- and Heavy-Duty Engines and Vehicles; Final Rules, 76 Fed. Reg. 57,106 (Sept. 15, 2011).

[58] Greenhouse Gas Emissions and Fuel Efficiency Standards for Medium- and Heavy-Duty; Final Rule, 81 Fed. Reg. 73,478 (Oct. 25, 2016). For more information on this rule, see CRS Insight IN10511, *Phase 2 Greenhouse Gas Emissions and Fuel Efficiency Standards for Heavy-Duty Vehicles*, by Richard K. Lattanzio.

In December 2016, a truck trailer group and racing vehicle coalition filed separate petitions for review of the Phase 2 GHG emission and fuel economy standards for medium- and heavy-duty engines and vehicles with the D.C. Circuit.[59] The Truck Trailer Manufacturers Association argued that EPA and NHTSA do not have authority to regulate GHG emissions and fuel economy of trailers because trailers are not "vehicles" under the CAA or the Energy Independence and Security Act.[60] The Racing Enthusiasts & Suppliers Coalition challenged EPA's prohibition on tampering and defeat devices, arguing that modifying vehicles for off-road racing or competition vehicles is beyond the scope of the CAA.[61] The court has granted motions to intervene from five environmental groups (Natural Resources Defense Council, the Environmental Defense Fund, Sierra Club, Center for Biological Diversity, and the Union of Concerned Scientists) and eight states (Connecticut, Washington, Massachusetts, Rhode Island, Vermont, Oregon, Iowa, and California, through its state Air Resources Board).[62]

California Preemption Waiver

In 1966, the California Motor Vehicle Pollution Control Board adopted auto emission standards for hydrocarbons and carbon monoxide, making California the first state to do so.[63] Decades later, in 2004, California adopted regulations requiring new motor vehicles to reduce GHG emissions, beginning in model year 2009.[64] While the CAA generally preempts states from adopting their own emission standards for mobile sources,[65] the CAA allows states with standards adopted prior to March 30,

---

[59] The cases have been consolidated under Truck Trailer Mfr. Ass'n, et. al. v. EPA, Nos. 16-1430, 16-1447 (D.C. Cir. Dec. 22, 2016).

[60] Petitioner's Non-Binding Statement of Issues at 2-3, Truck Trailer Mfr. Ass'n, et. al., v. EPA, No. 16-1430 (D.C. Cir. Feb. 22, 2017).

[61] Non-Binding Statement of Issues of Petitioner Racing Enthusiasts and Suppliers Coalition at 2, Racing Enthusiasts and Suppliers Coalition, v. EPA, No. 16-1447 (D.C. Cir. Feb. 22, 2017).

[62] Clerk's Order, Truck Trailer Manuf. Ass'n, et. al., v. EPA, No. 16-1430 (D.C. Cir. Mar. 10, 2017).

[63] Cal. Envtl. Prot. Agency, History of Air Resources Board, https://www.arb.ca.gov/knowzone/history.htm.

[64] Cal. Code Regs. tit. 13, §1961.1.

[65] 42 U.S.C. §7543(a).

1966, to obtain a waiver of preemption from EPA.[66] This waiver provision applies exclusively to California because it is the only state that had adopted standards within the prescribed timeframe. Other states, however, can adopt the California standards, provided that (1) the state has an EPA-approved State Implementation Plan; (2) the standards are identical to standards for which California has been granted a waiver; and (3) California and such state have adopted the standards at least two years before the commencement of the model year to which the standards apply.[67] Based on EPA's records, 14 states and the District of Columbia have adopted California standards.[68]

CAA Section 209(b) directs EPA to grant waivers unless certain factors warrant denial, providing that EPA "shall ... waive" the prohibition on state emission standards "if the State determines that the State standards will be, in the aggregate, at least as protective of public health and welfare as applicable Federal standards."[69] If California determines that its standards will be at least as protective as applicable federal standards, EPA can deny the waiver request only if EPA finds that California's determination is arbitrary and capricious (California is not required to show that its standards meet compelling and extraordinary conditions) or the California standards and accompanying enforcement procedures are inconsistent with CAA Section 202(a).[70] Legislative history indicates that when Congress amended the waiver provision in 1977, at least some Members of Congress intended for the provision to "afford California the broadest discretion possible in selecting the best means to protect the health of its citizens and the public welfare."[71]

EPA has granted in whole or in part numerous California waiver requests under Section 209(b). According to EPA records, California has

---

[66] *Id.* §7543(b).

[67] *Id.* §7507.

[68] *See* EPA, SALES OF CALIFORNIA VEHICLES FOR 2011 MODEL YEAR AND BEYOND (CROSS- BORDER SALES POLICY) Fig. 2 (2011) (listing state adoption of California standards as of Mar. 22, 2011), https://iaspub.epa.gov/otaqpub/ display_file.jsp?docid=24724&flag=1.

[69] 42 U.S.C. §7543(b).

[70] 42 U.S.C. §7543(b)(1)(A)-(C).

[71] H.R. Rep. No . 95-294, at 301-302 (May 12, 1977).

submitted over 100 waiver requests for new or amended standards or "within the scope" determinations (i.e., a request that EPA rule on whether a new regulation is within the scope of a waiver that the agency has already issued).[72] In 2013, EPA granted a waiver to California to regulate GHG emissions from light-duty vehicles through MY2025.[73] Most recently, in January 2017, the Obama Administration granted a series of waiver requests related to non-GHG emission standards for different types of mobile sources and engines.[74] If EPA revises its MTE determination and reduces the GHG standards for MY2022- 2025 light-duty vehicles,[75] California, and those states that have adopted California's standards, may potentially have standards different from the federal standards, resulting in a national patchwork of standards.

Aircraft

On August 15, 2016, and in response to various petitions,[76] EPA issued a finding that GHG emissions from commercial aircraft (1) contribute to

---

[72] *See* EPA, VEHICLE EMISSIONS CALIFORNIA WAIVERS AND AUTHORIZATIONS, https://www.epa.gov/state-and-localtransportation/vehicle-emissions-california-waivers-and-authorizations#state (listing Federal Register notices of waiver requests and decisions); Letter from Kevin de Leon, Pres. pro Tempore, 24 Cal. Senate District, et. al., to Xavier Becerra, Att'y Gen., Cal. Dep't of Justice (Mar. 16, 2017).

[73] California State Motor Vehicle Pollution Control Standards; Notice of Decision Granting a Waiver of Clean Air Act Preemption for California's Advanced Clean Car Program and a Within the Scope Confirmation for California's Zero Emission Vehicle Amendments for 2017 and Earlier Model Years, 78 Fed. Reg. 2,112 (Jan. 9, 2013).

[74] *See, e.g.*, California State Nonroad Engine Pollution Control Standards; Evaporative Emission Standards and Test Procedures for Off-Highway Recreational Vehicles (OHRVs); Notice of Decision, 82 Fed. Reg. 6,540 (Jan. 19, 2017); California State Motor Vehicle Pollution Control Standards; Amendments to On Highway Heavy-Duty Vehicle In-Use Compliance Program, Amendments to 2007 and Subsequent Model Year On-Highway Heavy-Duty Engines and Vehicles, and Amendments to Truck Requirements; Notice of Decision, 82 Fed. Reg. 4,867 (Jan. 17, 2017).

[75] *See supra* "Light-Duty Motor Vehicles" section.

[76] *See* Regulating Greenhouse Gas Emissions Under the Clean Air Act, 73 Fed. Reg. 44,354, 44,460 (July 30, 2008) (discussing the petitions asking EPA to make a finding that aircraft GHG emissions endanger public health or welfare, and that the agency adopt regulations that allow a range of compliance approaches, including emission limits, operational practices, fees, a cap-and-trade system, minimizing engine idling time, employing single engine taxiing, or use of ground-side electricity measures to replace the use of fuel-burning auxiliary power units at airport gates). Besides petitioning EPA for action on aviation emissions, environmental groups brought suit in the U.S. District Court for the District of Columbia seeking to force EPA to respond to their petitions on aircraft (as well as their petitions on marine vessels, and nonroad engines and vehicles). Ctr. for Biological Diversity

the pollution that causes climate change, and (2) endanger the health and welfare of Americans.[77] The finding, under Section 231 of the CAA, is the precondition for setting GHG emission standards for commercial aircraft. Aircraft manufacturers and airlines have not challenged EPA's finding.[78] It should be noted that the aircraft manufacturing and aviation industries have participated in international negotiations that resulted in an agreement on emission standards for new aircraft and a voluntary system to offset what might otherwise be a growth in emissions from air travel.[79]

Under the CAA, the EPA Administrator must consult with the Administrator of the Federal Aviation Administration and the Secretary of Transportation when developing emission standards, and may not impose new standards if doing so would significantly increase noise or adversely affect safety.[80] The President may also disapprove any such standards if the Secretary of Transportation finds that they would create a hazard to aircraft safety.[81]

v. U.S. EPA, 794 F.Supp. 2d 151 (D.D.C. 2011). On July 5, 2011, the court found that EPA has a mandatory duty to determine whether aircraft emissions endanger public health or welfare, and may be sued for unreasonable delay in doing so. *Id.* at 162. In March 2012, however, the court ruled that plaintiffs had not shown that EPA had unreasonably delayed such a decision. Ctr. for Biological Diversity v. EPA, No. 1:10-CV-985 (FJS), 2012 U.S. Dist. LEXIS 37870 (D.D.C. Mar. 20, 2012).

[77] 42 U.S.C. § 7571(a)(2). Finding That Greenhouse Gas Emissions from Aircraft Cause or Contribute to Air Pollution That May Reasonably Be Anticipated to Endanger Public Health and Welfare, 81 Fed. Reg. 54,422 (Aug. 15, 2016). For additional information on aircraft GHG emissions, see CRS Report R40090, *Aviation and Climate Change*, by James E. McCarthy.

[78] The finding was challenged by the Biogenic CO2 Coalition, and the case is in abeyance pending the resolution of other proceedings. *See* Biogenic CO2 Coalition v. EPA, D.C. Cir. No. 16-1358 (D.C. Cir. Oct. 14, 2016) (arguing that EPA did not distinguish the effects of biofuel CO2 emissions from the effects of similar emissions from fossil fuels). The coalition represents growers and processors of agricultural feedstocks that can be used for biofuels. THE BIOGENIC CO2 COALITION, http://www.biogenicco2.org/.

[79] For additional information on the international negotiations and agreement on aircraft GHG emissions and the ICAO process, see CRS Report R40506, *Cars, Trucks, Aircraft, and EPA Climate Regulations*, by James E. McCarthy and Richard K. Lattanzio, which notes: "In October 2016, the International Civil Aviation Organization (ICAO) agreed on international CO2 standards for aircraft, beginning in 2020. In October 2016, ICAO's governing council also agreed on a system for offsetting future carbon emissions from aviation—dubbed the Market-Based Mechanism, or MBM. The MBM agreed to in the ICAO process is voluntary for the next decade and has been agreed to by the U.S. industry." *Id.* at 13.

[80] 42 U.S.C. §7571(a)-(b).

[81] *Id.*

§7571(c).

Other Mobile Sources

CAA Section 213(a)(4), provides for EPA to promulgate emission standards for pollutants other than carbon monoxide (CO), oxides of nitrogen (NOx), and volatile organic compounds (VOCs)[82] from "nonroad engines and vehicles" (other than locomotives or their engines) if it makes an endangerment finding.[83] The endangerment language of Section 213 is similar to that for new motor vehicles in Section 202.[84] While CAA Section 213 does not set a deadline by which EPA must set standards for nonroad engines and vehicles, it provides for petitioners seeking GHG regulation of these mobile sources to file a suit against EPA for unreasonable delay in responding to rulemaking petitions.[85]

Following the *Massachusetts v. EPA* decision, EPA received petitions requesting EPA to regulate GHGs from mobile source categories.[86] These petitions covered aircraft, ocean-going ships and their fuels, motor fuels in general, locomotives, and nonroad vehicles and engines (construction equipment, farm equipment, logging equipment, outdoor power equipment, forklifts, marine vessels, recreational vehicles, and lawn and garden equipment).[87] In 2010, petitioners filed an "unreasonable delay" suit against EPA for failing to respond to three separate petitions submitted in

---

[82] *See* 42 U.S.C. §7547(a)(4) (referring to regulating pollutants other than "carbon monoxide, oxides of nitrogen, and volatile organic compounds" which are regulated under 42 U.S.C. §7547(a)(2)).

[83] 42 U.S.C. §7547(a)(4). The listed non-GHG pollutants are regulated under CAA §213(a)(3), which requires the imposition of best available control technology, and sets a deadline for such regulation.

[84] In place of the Section 209 words "cause, or contribute," Section 213 uses the phrase "significantly contribute": if the Administrator determines that emissions of GHGs from ships *significantly contribute* to air pollution which may reasonably be anticipated to endanger public health or welfare, he may promulgate such regulations as he deems "appropriate." *Id.* (emphasis added).

[85] Section 304(a)(3) of the Clean Air Act provides that "the district courts of the United States shall have jurisdiction to compel ... agency action unreasonably delayed," and requires that any person intending to file a legal action against the Administrator for unreasonable delay must provide notice of his or her intention to sue 180 days before commencing such action. 42 U.S.C. §7604; *see* 40 C.F.R. pt. 54. When notice of intent to sue is based on a failure to act, the notice must identify the provisions of the Clean Air Act that require the agency to take action, and describe the agency's failure to perform. 40 C.F.R. §54.3(a).

[86] 5 U.S.C. §553(e). For a list of these petitions, see Table 2 in CRS Report R40506, *Cars, Trucks, Aircraft, and EPA Climate Regulations*, by James E. McCarthy and Richard K. Lattanzio.

[87] *Id.*

2007 for GHG emissions rulemaking to cover marine vessels, nonroad vehicles, and aircraft engines.[88] In 2011, the federal district court ruled that EPA has discretion under Section 213 as to whether to issue endangerment findings for GHG emissions from marine vessels and nonroad vehicles.[89] The court explained that Section 213(a)(4) is "simply silent as to when—or whether— EPA must make endangerment findings; it merely says what EPA 'may' do 'if' an affirmative finding is made."[90] (The court ruled, however, that EPA does not have discretion under Section 231 over whether to make determinations regarding endangerment from aircraft GHG emissions.)[91] In 2012, EPA denied the petitions to regulate GHG and black carbon emissions from nonroad engines and vehicles, including marine vessels and engines.[92]

Executive Order Implications for GHG Regulation of Mobile Sources

The March 28, 2017, executive order did not specifically address mobile source GHG emission regulations or EPA's aircraft endangerment finding. While EPA and NHTSA could potentially identify these regulations or findings as rules that burden the "development or use" of energy resources that should be revised or rescinded, EPA and NHTSA would have to initiate a rulemaking to change the rules.

Stationary Sources under Clean Air Act

While mobile sources emit approximately 26% of total U.S. GHG emissions, stationary sources account for about 73% of the nation's GHG emissions.[93] According to EPA, electricity production generates the largest

---

[88] Complaint, Ctr. for Biological Diversity et al. v. EPA, No. 10-00985 (D.D.C. Jun. 6, 2010).

[89] Ctr. for Biological Diversity et al. v. EPA, 794 F. Supp. 2d 151, 157-158 (D.D.C. 2011).

[90] *Id.*

[91] *Id.* at 159-162.

[92] Memorandum in Response to Petitions Regarding Greenhouse Gas and other Emissions from Marine Vessels and Nonroad Engines and Vehicles (June 18, 2012), *available at* http://www.eenews.net/assets/2012/06/18/ document_pm_06.pdf.

[93] EPA, SOURCES OF GREENHOUSE GAS EMISSIONS, at https://www.epa.gov/ghgemissions/sources-greenhouse-gasemissions.

share of GHG emissions from stationary sources, producing 30% of 2014 domestic GHG emissions.[94]

Regulation of GHG Emissions under CAA Section 111

The Supreme Court's second decision on GHG emissions, *American Electric Power Co. v. Connecticut* (*AEP*),[95] unanimously held that EPA had authority to regulate GHGs under CAA Section 111(b) with respect to stationary sources. In *AEP*, the Court also held that the CAA displaced federal common law public nuisance claims related to GHG emissions from power plants because Congress had delegated to EPA the decision as to whether and how to regulate GHG emissions from power plants under CAA Section 111.[96]

New Source Performance Standards under Section 111(b)

CAA Section 111(b) directs EPA to establish standards of performance for listed categories of new, modified, or reconstructed stationary sources that "cause[], or contribute[] significantly to, air pollution which may reasonably be anticipated to endanger public health or welfare."[97] These standards are commonly referred to as new source performance standards or NSPSs. CAA Section 111 defines a "standard of performance" as an air pollution emission standard that reflects the "best system of emission reduction" (BSER) that has been "adequately demonstrated."[98] In

---

[94] *Id.*

[95] 564 U.S. 410 (2011).

[96] 564 U.S. at 424-26. For further discussion of *AEP*, see "Common Law Nuisance and Political Question Doctrine" section.

[97] 42 U.S.C. §7411(b)(1)(A). For example, EPA added "municipal solid waste landfills" as a Section 111 source category "because, in the judgement of the [EPA] Administrator, it contributes significantly to air pollution which may reasonably be anticipated to endanger public health and welfare." Standards of Performance for New Stationary Sources and Guidelines for Control of Existing Sources: Municipal Solid Waste Landfills, 61 Fed. Reg. 9,905 (March 12, 1996). *See also* Priority List and Additions to the List of Categories of Stationary Sources; Final Rule, 44 Fed. Reg. 49,222, 49,226 (Aug. 21, 1979) (listing "crude oil and natural gas production" as a source category under Section 111). Once EPA has promulgated new standards, states are allowed to develop and submit to EPA for approval a procedure for implementing and enforcing the standards of performance for new, modified, or reconstructed stationary sources. 42 U.S.C. §7411(b)(6). If the state procedure is approved, EPA will delegate implementation and enforcement authority to the state. *Id.*

[98] 42 U.S.C. §7411(a).

determining the BSER, EPA must take into account the cost of achieving the emission reductions; non-air quality health and environmental impacts; and the energy requirements of the regulated sources.[99]

Emission Guidelines under Section 111(d)

Coincident or after EPA issues CAA Section 111(b) standards for new, modified, and reconstructed sources, CAA Section 111(d) requires EPA to establish, under certain circumstances, standards of performance for *existing* sources in that category. Because Congress established a single definition for standards of performance promulgated pursuant to Sections 111(b) and 111(d), EPA must consider the same factors when establishing standards of performance for new, modified, reconstructed, and existing sources.[100]

CAA Section 111(d) directs EPA to establish state plan "procedures," which are similar to the CAA Section 110 guidelines, which EPA adopts to guide states in developing and revising implementation plans to achieve national ambient air quality standards (NAAQS).[101] EPA's

Section 111(d) emission guidelines establish binding requirements that states must address when they develop plans to implement standards of performance for existing sources in their jurisdictions.[102]

Power Plants

In 2011, EPA finalized a settlement agreement with states and others to promulgate NSPSs for GHG emissions from new, modified, and reconstructed fossil fuel-fired power plants under Section 111(b), and emission guidelines covering existing power plants under CAA Section

---

[99] *Id.*

[100] *Compare* 42 U.S.C. §7411(b)(1)(B) (requiring EPA to establish "standards of performance"), *with* 42 U.S.C. §7411(d)(1) (requiring EPA to establish a procedure "under which each State shall submit ... a plan which ... establishes standards of performance for any existing source"). The definition of "standards of performance" is provided under 42 U.S.C. §7411(a)(1), and applies to the provisions under Section 111.

[101] *Id.*; *see also* 42 U.S.C. §7410.

[102] 42 U.S.C. §7411(d)(1).

111(d).[103] In October 2015, EPA finalized the Section 111(b) NSPSs for GHGs from new power plants and Section 111(d) emission guidelines for GHGs from existing power plants.[104]

Both rules have been challenged in the D.C. Circuit; EPA's Section 111(d) emission guidelines rule, known as the Clean Power Plan (CPP), is stayed during the litigation.[105] On September 27, 2016, the en banc court heard oral arguments for the CPP,[106] but has not issued a decision yet. To resolve litigation challenging the Section 111(b) NSPSs for new power plants, a separate the three-judge panel has scheduled oral argument for April 17, 2017.[107]

In the March 28, 2017, executive order, President Trump directed EPA to take "immediate steps" to review the CPP and the GHG standards for new power plants, and, if the rules were inconsistent with the Administration's policy to promote coal, oil, natural gas, and nuclear energy production and use, to initiate a rulemaking to suspend, revise, or rescind the rules.[108] The executive order also requires EPA to review its proposed plan for states that choose not to develop their own plans to implement the CPP.[109] Under the executive order, EPA must notify the Attorney General of actions affecting the rules so that the Attorney General can take "appropriate" action in related litigation.[110] On March 28, 2017, the Department of Justice filed motions in both power plant cases, seeking

---

[103] *See* Settlement Agreement, New York v. EPA, No. 06-1322; and Am. Petroleum Inst. v. EPA, No. 08-1277 (D.C. Cir. Dec. 23, 2010), *available at* http://www2.epa.gov/sites/production/files/2013-09/documents/ boilerghgsettlement.pdf.

[104] Carbon Pollution Emission Guidelines for Existing Stationary Sources: Electric Utility Generating Units," Final Rule, 80 Fed. Reg. 64,661 (Oct. 23, 2015) [hereinafter *Clean Power Plan*]; Standards of Performance for Greenhouse Gas Emissions From New, Modified, and Reconstructed Stationary Sources: Electric Utility Generating Units," Final Rule, 80 Fed. Reg. 64,510 (Oct. 23, 2015) [hereinafter *New Power Plant NSPS*].

[105] For more information on the litigation and the stay, see CRS Report R44480, *Clean Power Plan: Legal Background and Pending Litigation in West Virginia v. EPA*, by Linda Tsang and Alexandra M. Wyatt.

[106] *See id.*; *see also* Ellen M. Gilmer, *Clean Power Plan: Oral Arguments: The Blow by Blow*, E&E NEWS, September 28, 2016, http://www.eenews.net/stories/1060043538.

[107] Order, North Dakota, et. al., v. EPA, No. 15-1381 (D.C. Cir. Dec. 2, 2016).

[108] Exec. Order, Promoting Energy Independence and Economic Growth §4 (Mar. 28, 2017).

[109] *Id.* §4(b)(iii).

[110] *Id.* §4(d).

to hold the cases in abeyance until 30 days after EPA completes its review of the rules and subsequent rulemakings resulting from its review.[111]

The executive order also directs EPA to review and potentially revise or rescind the legal memorandum that EPA released with the final rule.[112] It is possible that EPA could revise its interpretation of two different versions of a 1990 amendment to CAA Section 111(d) to accord with the view that regulation of hazardous air pollutant from power plants under CAA Section 112 bars EPA's regulation of CO2 emissions from power plants under CAA Section 111(d).[113] It is also possible that, through a new rulemaking, EPA could rescind the NSPS for new power plants, which is a prerequisite for EPA's authority under Section 111(d) to regulate existing power plants.[114] Administrator Pruitt has also mentioned that EPA may reconsider its 2009 endangerment finding[115] that underlies almost all CAA GHG regulation.[116] All of these options would effectively remove EPA's authority to implement the CPP or issue any Section 111(d) emission guidelines for existing power plants.

EPA could also revise the stringency of the GHG standards and emissions guidelines for new power plants and focus the emission guidelines on what GHG reductions for existing power plants can achieve "inside the fenceline."[117] Exercising these options would affect issues

---

[111] Notice of Exec. Order, EPA Review of Clean Power Plan and Forthcoming Rulemaking, and Motion to Hold Cases in Abeyance, West Virginia, et. al. v. EPA, No. 15-1363 (D.C. Cir. Mar. 28, 2017); Notice of Exec. Order, EPA Review of Rule and Forthcoming Rulemaking, and Motion to Hold Cases in Abeyance, North Dakota, et. al. v. EPA, No. 15-1381 (D.C. Cir. Mar. 28, 2017).

[112] Exec. Order, Promoting Energy Independence and Economic Growth §4(c) (Mar. 28, 2017).

[113] See discussion of the Section 112 Exclusion in CRS Report R44480, *Clean Power Plan: Legal Background and Pending Litigation in West Virginia v. EPA*, by Linda Tsang and Alexandra M. Wyatt.

[114] *See id.*, "Clean Air Act Section 111" section.

[115] *See supra* "Reducing GHG Emissions" section.

[116] Coral Davenport, *E.P.A. Chief Doubts Consensus View of Climate Change*, N.Y. Times, Mar. 9, 2017. One stakeholder has filed a petition for reconsideration of the 2009 endangerment finding. *See* Petition of the Competitive Enterprise Institute and the Science and Environmental Policy Project for Rulemaking on the Subject of Greenhouse Gases and Their Impact on Public Health and Welfare, in Connection with EPA's 2009 Endangerment Finding, 74 FR 66,496 (Dec. 15, 2009) (filed on Feb. 17, 2017).

[117] Petitioners challenging the CPP have argued that Section 111 authorizes EPA to require only measures that can be applied to the "performance" of an individual "source" (also known as measures "inside the fenceline"), such as adoption of pollution control devices or other

central to litigation over the rules. Petitioners have argued that EPA exceeded its authority in basing the best system for emission reduction for new power plants on carbon capture and sequestration technology and going beyond the fenceline of existing power plants to achieve greater GHG reductions.[118] Further, retaining some version of the GHG power plant rules would support that EPA has fulfilled its duty to promulgate GHG regulations for power plants pursuant to the judicial settlement agreement that initiated, in part, the promulgation of these rules.[119]

As discussed above, for EPA to pursue any of these actions, the agency would have to go through a new rulemaking and would be required to provide a "reasonable justification" for reversing course on the legal theories that supported the rules or reducing the stringency of the GHG standards.[120]

Refineries

Petroleum refineries have been subject to NSPSs for a variety of pollutants since the CAA was passed in 1970.[121] In 2010, EPA entered into a judicial settlement that required EPA to issue regulations addressing GHGs from petroleum refineries.[122] The deadline for a final refinery rule

design or operational standards. *See generally* Opening Br. of Pet'rs on Core Legal Issues, at 29-61, West Virginia, et al. v. EPA, No. 15-1363 (D.C. Cir. Feb. 19, 2016).

[118] See "Section 111(d) Scope of Authority" and "North Dakota v. EPA: Section 111(b) Litigation" sections, CRS Report R44480, *Clean Power Plan: Legal Background and Pending Litigation in West Virginia v. EPA*, by Linda Tsang and Alexandra M. Wyatt.

[119] *See* Settlement Agreement, *supra* note 103.

[120] *See* CRS Legal Sidebar WSLG888, *How to Repeal a Rule*, by Jared P. Cole, The Federal Rulemaking Process: An Overview, coordinated by Maeve P. Carey.

[121] *See* Standards of Performance for Electric Utility Sim Generating Units, 40 C.F.R. pt. 60, subpt. Da (specifying rate-based performance standards or percent reduction standards for particulate matter, sulfur dioxide, and nitrogen oxides from power plants built after certain dates); National Emission Standards for Hazardous Air Pollutants: Coal- and Oil-Fired Electric Utility Steam Generating Units, 40 C.F.R. pt. 63, subpt. UUUUU (specifying standards to limit mercury, acid gases and other hazardous air pollution from power plants).

[122] *See* Settlement Agreement, New York v. EPA, No. 06-1322; and Am. Petroleum Inst. v. EPA, No. 08-1277 (D.C. Cir. Dec. 23, 2010), *available at* https://www.epa.gov/sites/production/files/2013-09/documents/refineryghgsettlement.pdf [hereinafter *Refinery Settlement Agreement*].

identified in the settlement was originally November 10, 2012.[123] At EPA's request, this deadline was extended four times.[124]

During these delays, EPA was working to finalize the NSPSs and emission guidelines to address CO2 emissions from power plants, and methane emissions from oil and gas sources.[125] In the most recent status report filed with the court, EPA stated that it is "working diligently" on a rule and will continue to assess how best to resolve the remaining obligations (including the promulgation of a GHG rule for refineries) from the settlement agreement.[126] As part of the agreement, the environmental groups and states have retained the right to sue the EPA if the agency does not meet its obligations.[127]

Regulation of Methane Emissions from Stationary Sources

One of the initiatives in the Obama Administration's CAP focused on controlling methane emissions, a short-lived climate pollutant with a potential Global Warming Potential[128] of more than 25 times that of CO2.[129] EPA and other federal agencies developed an interagency "Strategy to Reduce Methane Emissions" (Methane Strategy) that outlined voluntary actions and potential agency rulemakings to cut methane emissions.[130] To implement the strategy, EPA focused on updating and

---

[123] *Id.* at 4.

[124] Maria Chutchian, *EPA Will Miss 2012 Deadline For GHG Refinery Rule*, LAW360, Feb. 29, 2012, https://www.law360.com/articles/314527.

[125] *See supra* "Power Plants" and "Regulation of Methane Emissions from Stationary Sources" sections.

[126] Status Report, Am. Petroleum Inst. v. EPA, No. 08-1277 (D.C. Cir. Feb. 14, 2017).

[127] *Refinery Settlement Agreement*, *supra* footnote 122, at 6.

[128] The Global Warming Potential (GWP) is a metric adopted by EPA to compare the climate impacts of different gases. EPA, *Understanding Global Warming Potentials*, https://www.epa.gov/ghgemissions/understanding-globalwarming-potentials. The GWP measures the total energy the emissions of 1 ton of a gas will absorb over a given period of time, relative to the emissions of 1 ton of CO2. *Id.*

[129] Methane (CH4) is estimated to have a GWP of 28–36 over 100 years. *Id.* For additional information on the GWP of methane, see CRS Report R43860, *Methane: An Introduction to Emission Sources and Reduction Strategies*, coordinated by Richard K. Lattanzio.

[130] Exec. Off. of the President, Climate Action Plan: Strategy to Reduce Methane Emissions 1-2 (2014), *available at* https://obamawhitehouse.archives.gov/blog/2014/03/28/strategy-cut-methane-emissions [hereinafter *Methane Strategy*]. For additional information and background on methane and the Strategy, see CRS Report R43860, *Methane: An*

revising NSPSs and emissions guidelines to reduce methane emissions from new and existing municipal landfills and sources in the oil and gas sector.[131] The Department of the Interior's Bureau of Land Management (BLM) released a rule to address the venting and flaring of methane gas on federal and Indian lands.[132] This section will review these rules and related legal challenges and pending congressional actions.

Municipal Landfills

In response to a consent decree resolving a lawsuit[133] and the Obama Administration's goal to reduce methane under the CAP,[134] EPA proposed updates to the 1996 municipal solid waste (MSW) landfill NSPSs and emission guidelines in 2014.[135] Over 20 years after the original NSPSs and emission guidelines were issued, EPA published on August 29, 2016, its updated NSPSs to reduce landfill gas emissions from MSW landfills built,

---

Introduction to Emission Sources and Reduction Strategies, coordinated by Richard K. Lattanzio.

131 Standards of Performance for Municipal Solid Waste Landfills, Final Rule, 81 Fed. Reg. 59,332 (Aug. 29, 2016) (codified at 40 C.F.R. pt. 60, subpt. XXX) [hereinafter *2016 Landfill NSPS*]; Emission Guidelines and Compliance Times for Municipal Solid Waste Landfills; Final Rule, 81 Fed. Reg. 59,276 (Aug. 29, 2016) (codified at 40 C.F.R. pt. 60, subpt. WWW); Oil and Natural Gas Sector: Emission Standards for New, Reconstructed, and Modified Sources, Final Rule, 81 Fed. Reg. 35,824, at 35,840, 35,848 (June 3, 2016) (to be codified at 40 C.F.R. pt. 60, subpts. OOOO and OOOOa) [hereinafter "2016 Oil & Gas NSPS"].

132 Waste Prevention, Production Subject to Royalties, and Resource Conservation, 81 Fed. Reg. 83,008 (Nov. 18, 2016).

133 Environmental Defense Fund filed a lawsuit against EPA for failure to review the 1996 NSPS by the statutorily required deadline. Envt'l Def. Fund v. EPA. No. 1:11-cv-04492 (S.D.N.Y. June 30. 2011). Under a consent decree resolving that lawsuit, EPA agreed to review the 1996 landfill NSPS and take final action on whether to revise them. Consent Decree, Envt'l Def. Fund v. EPA. No. 1:11-cv-04492 (S.D.N.Y. October 1, 2012).

134 *Methane Strategy*, *supra* note 130. For additional information and background on methane and the Strategy, see CRS Report R43860, *Methane: An Introduction to Emission Sources and Reduction Strategies*, coordinated by Richard K. Lattanzio.

135 In 1996, EPA listed MSW landfills as a CAA Section 111 source category, and issued NSPSs for new, modified, and reconstructed MSW landfills pursuant to Section 111(b) and emission guidelines for existing landfills pursuant to Section 111(d) to address nonmethane organic compounds (NMOCs) and methane emissions. Standards of Performance for New Stationary Sources and Guidelines for Control of Existing Sources: Municipal Solid Waste Landfills, 61 Fed. Reg. 9,905 (March 12, 1996). For additional information on previous MSW emission standards, see CRS Report R43860, *Methane: An Introduction to Emission Sources and Reduction Strategies*, coordinated by Richard K. Lattanzio, and CRS Report R44615, *EPA's Recent Methane Regulations: Legal Overview*, by Linda Tsang.

modified, or reconstructed after July 17, 2014.[136] EPA also revised emission guidelines for existing landfills operating prior to that date.[137] Both rules became effective on October 28, 2016.[138] While the March 28, 2017, executive order did not address specifically the MSW landfill rules, EPA could potentially identify these regulations as burdening the "development or use" of energy resources and recommend their revision or rescission.[139] EPA would then have to initiate a rulemaking to do so.[140]

Industry trade associations and waste management and recycling companies have challenged EPA's 2016 revised emission guidelines for existing MSW landfills in the D.C. Circuit.[141] The court has not set a briefing schedule yet. Petitioners challenging the landfill emissions guidelines question whether EPA has the authority to revise the 1996 emission guidelines for existing MSW landfills under CAA Section 111(d).[142] Pursuant to CAA Section 111(b), EPA is required "at least every 8 years [to] review and, if appropriate, revise" NSPSs for new, modified, and reconstructed sources.[143] However, Section 111(d) does not include a similar mandated review period for emission guidelines for *existing* sources.[144] EPA's implementing regulations for Section 111(d) emission guidelines only address revisions to the agency's determination regarding adverse public health effects that may alter compliance times under the

---

[136] See 2016 Landfill NSPS. For additional information on landfill standards, see CRS Report R43860, *Methane: An Introduction to Emission Sources and Reduction Strategies*, coordinated by Richard K. Lattanzio, and CRS Report R44615, *EPA's Recent Methane Regulations: Legal Overview*, by Linda Tsang.

[137] Emission Guidelines and Compliance Times for Municipal Solid Waste Landfills; Final Rule, 81 Fed. Reg. 59,276 (Aug. 29, 2016) (codified at 40 C.F.R. pt. 60, subpt. WWW).

[138] *Id.*; *2016 Landfill NSPS*, at 59,322.

[139] Exec. Order, Promoting Energy Independence and Economic Growth §2 (Mar. 28, 2017).

[140] *Id.* §2(g).

[141] Nat'l Waste & Recycling Ass'n v. EPA, No. 16-1371 (D.C. Cir., filed Oct. 27, 2016); Utility Air Regulatory Group v. EPA, No. 16-1374 (D.C. Cir., filed Oct. 28, 2016).

[142] Nonbinding Statement of Issues of Petition Utility Air Reg. Group, No. 16-1374 (consolidated with No. 16-1371) (D.C. Cir. Dec. 1, 2016). For addition information this legal argument, see CRS Report R44615, *EPA's Recent Methane Regulations: Legal Overview*, by Linda Tsang.

[143] 42 U.S.C. §7411(b)(1)(B).

[144] *See id.* §7411(d).

guidelines and require states to make corresponding revisions to their state plans.[145]

A ruling upholding EPA's implicit authority to review and revise Section 111(d) guidelines would support future EPA review and revision of other Section 111(d) emissions guidelines to require additional emission reductions based on a new analysis of the "best system of emission reduction."[146] For example, if EPA follows the Section 111(b) eight-year review cycle, EPA could consider revisions of the CPP (if upheld) in 2023—one year after the CPP's first mandatory emission reduction period starts.[147]

Sources from the Oil and Gas Sector

In 2012, EPA issued a final rule setting standards of performance to limit emissions of VOCs from new, reconstructed, and modified sources in the oil and gas industry pursuant to CAA Section 111(b) (2012 NSPSs).[148] In response to the Methane Strategy and petitions for reconsideration of the

---

[145] *See* 40 C.F.R. §60.22(d) (requiring EPA to "publish notice of the determination in the Federal Register, revise the guideline document as necessary, and propose and promulgate emission guidelines and compliance times" if the agency concludes that on the basis of new information that a prior determination regarding adverse health effects of a designated pollutant is incorrect or no longer correct).

[146] In addition to the Clean Power Plan and the 1996 guidelines for landfills, EPA has issued Section 111(d) emission guidelines to address acid mist from sulfuric acid production units; fluoride emissions from phosphate fertilizer plants; reduced sulfur emissions from kraft pulp mills; and fluoride emissions from primary aluminum plants. *See* 80 Fed. Reg. 64,662 (October 23, 2015); 61 Fed. Reg. 9,905 (Mar. 12, 1996); 60 Fed. Reg. 65,387 (December 19, 1995); 45 Fed. Reg. 26,294 (Apr. 17, 1980); 44 Fed. Reg. 29,828 (May 22, 1979); 42 Fed. Reg. 12,022 (Mar. 1, 1977). In addition, EPA's 2005 Clean Air Mercury Rule (CAMR) delisted coal-fired power plants from CAA Section 112 and, instead, established a cap-and-trade system for mercury under Section 111(d). 70 Fed. Reg. 28,606 (May 18, 2005). The D.C. Circuit Court vacated CAMR in 2008 on grounds unrelated to the guidelines' substantive requirements. *See* New Jersey v. EPA, 517 F.3d 574, 581-84 (D.C. Cir. 2008) (holding that EPA's delisting of the source category from Section 112 was unlawful and that EPA was obligated to promulgate standards for mercury and other hazardous air pollutants under § 112).

[147] Carbon Pollution Emission Guidelines for Existing Stationary Sources: Electric Utility Generating Units; Final Rule, 80 Fed. Reg. 64,662, 64,673 (October 23, 2015).

[148] Oil and Natural Gas Sector: New Source Performance Standards and National Emission Standards for Hazardous Air Pollutants Reviews; Final Rule, 77 Fed. Reg. 49,542 (Aug. 16, 2012) (codified at 40 C.F.R. pt. 60, subpt. OOOO).

2012 NSPSs to address methane emissions,[149] EPA issued a rule revising and updating the 2012 NSPSs on June 3, 2016.[150] EPA's final rule amends the 2012 NSPS, expanding the emission sources covered by the rule, and establishes performance standards for GHGs (in the form of methane limitations) from a number of oil and gas emission sources.[151] The March 28, 2017, executive order requires EPA to review this rule, and if inconsistent with the Administration's policy to promote coal, oil, natural gas, and nuclear energy production and use, initiate a rulemaking to suspend, revise, or rescind the rule.[152]

North Dakota,[153] Texas (including the Railroad Commission of Texas and the Texas Commission on Environmental Quality as named parties),[154] a coalition of 14 states (including several state agencies),[155] and various gas associations filed petitions for review of the final rule.[156] Nine states and the city of Chicago moved to intervene on behalf of EPA to support the

---

[149] *See e.g.*, Clean Air Council et al., Petition for Reconsideration, In the Matter of: Final Rule Published at 77 FR 49490 (Aug. 16, 2012).

[150] Oil and Natural Gas Sector: Emission Standards for New, Reconstructed, and Modified Sources, Final Rule, 81 Fed. Reg. 35,824, at 35,840, 35,848 (June 3, 2016) (to be codified at 40 C.F.R. pt. 60, subpts. OOOO and OOOOa) (hereinafter "2016 Oil & Gas NSPS"); 42 U.S.C. §7411. The rule took effect on Aug. 2, 2016, and remains in effect during litigation unless the petitioners seek and are granted a stay of implementation of the rule by the court. *Id.* at 35,824.

[151] *Id.* at 35,825. For additional information on the requirements of the rule, see CRS Report R43860, *Methane: An Introduction to Emission Sources and Reduction Strategies*, coordinated by Richard K. Lattanzio, and CRS Report R44615, *EPA's Recent Methane Regulations: Legal Overview*, by Linda Tsang.

[152] Exec. Order, Promoting Energy Independence and Economic Growth §7 (Mar. 28, 2017).

[153] Petition for Review, North Dakota v. EPA, No. 16-1242 (D.C. Cir. July 15, 2016).

[154] Petition for Review, Texas et al. v. EPA, No. 16-1257 (D.C. Cir. July 28, 2016).

[155] Petition for Review, West Virginia et al. v. EPA, No. 16-1264 (D.C. Cir. Aug. 2, 2016). Besides West Virginia, the state coalition includes Alabama, Arizona, Kansas, Kentucky, Louisiana, Att'y Gen. Bill Schuette for the people of Michigan, Montana, Ohio, Oklahoma, South Carolina, Wisconsin, Kentucky Energy and Environment Cabinet, and the North Carolina Department of Environmental Quality. *Id.* On March 20, 2017, the court granted North Carolina Department of Environmental Quality's motion to withdraw as a petitioner. Clerk's Order, West Virginia et al. v. EPA, No. 16-1264 (D.C. Cir. Mar. 20, 2016).

[156] Petition for Review, W. Energy Alliance v. EPA, No. 16-1266 (D.C. Cir. Aug. 2, 2016); Petition for Review, Indep. Petroleum Ass'n of Am. et al. v. EPA, No. 16-1262 (D.C. Cir. Aug. 2, 2016); Petition for Review, Interstate Nat. Gas Ass'n, No. 16-1263 (D.C. Cir. Aug. 2, 2016); Petition for Review, GPA Midstream Gas Ass'n, No. 16-1267 (D.C. Cir. Aug. 2, 2016); Petition for Review, Texas Oil & Gas Ass'n, No. 16-1269 (D.C. Cir. Aug. 2, 2016); Petition for Review, Am. Petroleum Inst., No. 16-1270 (D.C. Cir. Aug. 2, 2016).

final rule.[157] In addition, environmental advocacy groups filed their own motion to intervene in the case.[158] All petitions have been consolidated with the lead case, *North Dakota v. EPA*.[159]

Based on the preliminary statements of issues filed, the petitioners claim, among other things, that EPA exceeded its statutory authority; failed to make the required "endangerment finding" for methane from the oil and gas sector under CAA Section 111; and "unlawfully" expanded the listed oil and natural gas sector source category to additional types of emission sources not previously regulated.[160] The final litigated issues will be presented in the petitioners' briefs, which have not been filed yet.[161] On April 7, 2017, the Department of Justice filed a motion, seeking to hold the cases in abeyance until 30 days after EPA completes its review of the rules and subsequent rulemakings resulting from its review.[162]

In 2016, EPA sent an Information Collection Request (ICR) to oil and natural gas companies seeking information on their existing oil and gas sources as a first step to regulating their methane emissions.[163] Effective March 2, 2017, EPA withdrew the ICR to assess the need for this information and to reduce the burden to businesses during this assessment.[164] The ICR withdrawal was issued shortly after nine state

---

[157] Unopposed Mot. California, Connecticut, Illinois, New Mexico, New York, Oregon, Rhode Island, Vermont, Massachusetts, and the City of Chicago for Leave to Intervene as Resp'ts, North Dakota v. EPA, No. 16-1242 (D.C. Cir. Aug. 15, 2016).

[158] Nat. Res. Def. Council, Envtl. Def. Fund, Sierra Club, Clean Air Council, Earthworks, and Envtl. Integrity Project Motion to Intervene as Resp'ts, North Dakota v. EPA, No. 16-1242 (D.C. Cir. Aug. 15, 2016).

[159] *See* Clerk's Order, North Dakota v. EPA, No. 16-1242 (D.C. Cir. Aug. 8, 2016).

[160] *See* Petitioner N.D.'s Statement of Issues to be Raised, North Dakota v. EPA, No. 16-1242 (D.C. Cir. Aug. 22, 2016); Petitioner W. Energy Alliance Non-Binding Statement of Issues to be Raised, North Dakota v. EPA, No. 16 1242 (D.C. Cir. Sept. 1, 2016).

[161] D.C. CIR. R. 28(a)(5).

[162] Notice of Exec. Order and Motion to Hold Cases in Abeyance, Am. Petroleum Inst. v. EPA, No. 13-1108 (D.C. Cir. Apr. 7, 2017).

[163] Proposed Information Collection Request; Comment Request; Information Collection Effort for Oil and Gas Facilities, 81 Fed. Reg. 35,763 (June 3, 2016).

[164] Notice Regarding Withdrawal of Obligation To Submit Information, 82 Fed. Reg. 12,817 (Mar. 7, 2017).

attorneys general and two governors submitted a letter to EPA, asking that the ICR be suspended and withdrawn.[165]

Venting and Flaring of Methane Gas

On November 18, 2016, BLM promulgated new regulations to reduce the loss of natural gas (and methane which is a primary component of natural gas) from venting, flaring, and leaks during oil and natural gas production activities on onshore federal and Indian leases.[166] These regulations replace the existing provisions related to venting, flaring, and royalty-free use of gas issued 30 years prior under the Mineral Leasing Act of 1920.[167] The rule requires owners and operators of both new and existing federal and Indian oil and gas leases to implement comprehensive methane leak detection and repair (LDAR) programs throughout their extraction, storage and transmission infrastructures, and will also place new restrictions on the venting and flaring of methane during drilling and storage operations on those leases by establishing stringent new methane emissions reduction and capture targets.[168] By January 17, 2018, lessees will be required to recover or flare 85% of all methane produced on federal and Indian oil and gas leases.[169] That number increases to 95% in 2020, and to 98% beginning in 2026.[170]

The rule was immediately challenged. On November 28, 2016, the Western Energy Alliance, Independent Petroleum Association of America, and the States of Wyoming, Montana, and North Dakota filed petitions for review with the U.S. District Court in the District of Wyoming.[171] The

---

[165] Letter from Ken Paxton, Att'y. Gen., Texas, to Scott Pruitt, Admin., EPA (Mar. 1, 2017), *available at* https://www.epa.gov/sites/production/files/2017-03/documents/letter_from_attorneys_general_and_governors.pdf.

[166] Waste Prevention, Production Subject to Royalties, and Resource Conservation, 81 Fed. Reg. 83,008 (Nov. 18, 2016).

[167] 43 C.F.R. pt. 3160.

[168] 81 Fed. Reg. at 83,010.

[169] *Id.* at 83,011.

[170] *Id.*

[171] Wyoming v. Dep't. of Interior, No. 2:16-cv-00285 (D. Wyo., filed Nov. 15, 2016) (lead case in the consolidated cases). The court denied petitioners' motion for a preliminary injunction to block the rule pending resolution of the litigation. Orders on Motions for Preliminary Injunction, Wyoming v. Dep't. of Interior, No 2:16-cv-00280 (D. Wyo. Jan. 16, 2017).

petitioners claim, among other things, that BLM exceeded its statutory authority by attempting to regulate atmospheric emissions of methane—an authority the petitioners argue that Congress explicitly delegated to the states and the EPA.[172] On March 21, 2017, Texas filed a motion to intervene as a petitioner in the lawsuit.[173] Since the court has denied the petitioners' motion for a preliminary injunction, the rule took effect on January 17, 2017.[174]

Since the initial petition was filed, 15 environmental advocacy groups and NGOs, in addition to the States of California and New Mexico, have intervened on behalf of BLM in support of the new rule. Supporters of the rule and BLM argue that Congress delegated authority to BLM to consider environmental impacts in the promulgation of rules and regulations regarding the use of federal lands.[175]

As the litigation continues, Congress is taking action to repeal the BLM venting and flaring rule. On February 3, 2017, the House voted 221-191 to approve H.J.Res. 36, a Congressional Review Act (CRA) resolution to disapprove the rule.[176] The resolution is now being considered in the Senate, where Environment and Public Works Chairman John Barrasso has introduced a separate CRA resolution of disapproval, S.J.Res. 11.[177] Supporters of the rule claim that it will reduce methane pollution, and put

---

http://wordpress2.ei.columbia.edu/climate-change-litigation/files/case-documents/2017/20170116_docket-216-cv-00285_order.pdf?mc_cid=32fd774ef0&mc_eid=53146628a6.

172 Orders on Motions for Preliminary Injunction, Wyoming v. Dep't. of Interior, No 2:16-cv-00280 (D. Wyo. Jan. 16, 2017).

173 Unopposed Motion to Intervene as Petitioner for Review of Final Agency Action, No 2:16-cv-00280 (D. Wyo. Mar. 21, 2017).

174 Waste Prevention, Production Subject to Royalties, and Resource Conservation, 81 Fed. Reg. at 83,008.

175 See Federal Respondents' Consolidated Opposition to Petitioners' and Petitioner-Intervenor's Motions for Preliminary Injunction & Citizen Groups' Response to Motions for a Preliminary Injunction, No 2:16-cv-00280 (D. Wyo. Dec. 15, 2016).

176 H.J.Res. 36, 115th Cong. (2017). For more information regarding the Congressional Review Act, see CRS Report R43992, The Congressional Review Act (CRA): Frequently Asked Questions, by Maeve P. Carey and Christopher M. Davis; CRS In Focus IF10023, The Congressional Review Act (CRA), by Maeve P. Carey and Christopher M. Davis; and CRS Insight IN10660, What Is the Effect of Enacting a Congressional Review Act Resolution of Disapproval?, by Maeve P. Carey.

177 S.J.Res. 11, 115th Cong. (2017).

more natural gas into the energy market by capturing the gas.[178] Opponents of the rule argue that it restricts the industry, affects employment, and potentially raises natural gas prices.[179] Industry groups claim the rule will "further impede oil and natural gas production on federal land, which already has been declining" and is redundant of voluntary efforts and existing state regulations to reduce methane gas releases.[180]

The March 28, 2017, executive order requires BLM to review this rule, and if inconsistent with the Administration's policy to promote coal, oil, natural gas, and nuclear energy production and use, initiate a rulemaking to suspend, revise, or rescind the rule.[181]

Regulation of GHG Emissions from Stationary Sources through Permits

EPA interpreted the mobile source GHG regulations under CAA Section 202 as triggering regulations under the Prevention of Significant Deterioration (PSD) program[182] and Title V permitting program.[183] The PSD program imposes requirements in addition to those in Section 111 for new or modified stationary emission sources in areas that have met or exceeded the national ambient air quality standards (NAAQSs).[184] This program requires preconstruction permits for new major emitting facilities and new major modifications.[185] PSD permits contain emission limits based on the best available control technology (BACT) for each regulated pollutant emitted from that source.[186]

---

[178] See Devin Henry, House Votes to Overturn Obama Drilling Rule, THE HILL, Feb. 3, 2017, http://thehill.com/policy/energy-environment/317739-house-votes-to-overturn-obama-oil-and-gas-rule; Pamela King, Westerners Fear GOP's'morally outrageous' BLM Rule Rollback, E&E NEWS, Feb. 3, 2017, http://www.eenews.net/stories/1060049488.

[179] See Rep. Kevin McCarthy, BLM Methane Venting and Flaring Rule, https://www.majorityleader.gov/2017/02/02/blm-methane-venting-flaring-rule/.

[180] Devin Henry, House Votes to Overturn Obama Drilling Rule, THE HILL, Feb. 3, 2017.

[181] Exec. Order, Promoting Energy Independence and Economic Growth §7(b)(iv) (Mar. 28, 2017).

[182] 42 U.S.C. §§7470-7479.

[183] See id. §§7661-7661f, 7602(j).

[184] Id. §§7475-7479.

[185] Id. §7475(a)(4).

[186] *Id.* §7475(a)(4).

The Title V program requires major sources of air pollutants and certain other sources to obtain and operate in compliance with an operating permit, commonly referred to as Title V operating permits.[187] These permits include pollution control requirements from federal or state regulations that apply to the source, as well as other applicable air permits such as a PSD permit.[188] Sources with these Title V permits are required by the act to certify compliance with the applicable requirements of their permits at least annually.[189]

The Supreme Court's third major decision on GHG emissions, *Utility Air Regulatory Group v. EPA* (*UARG*), rejected EPA's argument that its regulation of motor vehicle GHG emissions required it to regulate GHG emissions from stationary sources under the PSD and Title V programs.[190] The Court held that EPA's regulation of vehicle GHG emissions does *not* give EPA unqualified authority to require PSD and Title V permitting programs to stationary source GHG emissions.[191]

As the Court explained, just because the phrase "air pollutant" encompasses GHGs in some provisions of the CAA, which it had held in *Massachusetts v. EPA*, the phrase "air pollutants" does not necessarily include GHGs in all the CAA provisions, such as in the PSD and Title V sections.[192] Reading "air pollutant" to include GHG emissions in these permit programs, the Court reasoned, creates a staggering administrative workload because of the low emission thresholds that trigger those programs, and the huge number of sources that satisfy those thresholds for CO2, the primary GHG.[193] The administrative unwieldiness of demanding PSD and Title V permits for so many sources argued strongly, in the

---

[187] *Id.* Detailed requirements of the Title V permitting program are set forth at 40 C.F.R. pt. 70 for state-issued Title V permits and pt. 71 for federally-issued permits. For more information about the Clean Air Act, see CRS Report RL30853, *Clean Air Act: A Summary of the Act and Its Major Requirements*, by James E. McCarthy and Claudia Copeland.

[188] *Id.* §7661c(f).

[189] 40 C.F.R. §70.5.

[190] Util. Air Regulatory Group [UARG] v. EPA,—U.S.—, 134 S. Ct. 2427, 2439-46 (2014).

[191] *Id.*

[192] *Id.* at 2439-32.

[193] *Id.*

Court's view, against a GHG-inclusive reading of those programs.[194] Nor did the Court allow EPA, through its "Tailoring Rule,"[195] to raise the low statutory emission thresholds to 75,000 tons per year (tpy) CO2-equivalents (CO2-e) in an effort to ease the permit-issuing workload, since the CAA states the thresholds in absolute numerical terms.[196]

The Court allowed, however, that EPA could impose one aspect of the PSD program with respect to GHGs. When PSD permitting is required because a new (or modified) source emits a conventional" non-GHG pollutant at or above threshold quantities, the Court concluded that EPA also may impose the BACT requirement of the PSD program on these "anyway" sources that emitted more than a *de minimis* amount of GHG emissions.[197]

The narrowness of the question that the Court answered in *UARG* is significant, since it leaves intact the D.C. Circuit's approval of EPA's endangerment finding for GHG emissions from new motor vehicles and the *Massachusetts v. EPA* holding. Nor did *UARG* directly touch on the validity of EPA's current use of CAA Section 111 in its effort to control GHG emissions from new and existing fossil-fuel-fired power plants.[198]

---

[194] *Id.* at 2442-44. "[T]he PSD program and Title V are designed to apply to, and cannot be extended beyond, a relative handful of large sources capable of shouldering heavy substantive and procedural burdens." *Id.* at 2443.

[195] On May 13, 2010, EPA finalized its "tailoring" rule aimed to relieve the overwhelming permitting burdens that EPA asserted would, in the absence of the rule, fall on PSD and Title V permitting authorities beginning January 2, 2011 the effective date of the light-duty vehicle rule discussed above. Prevention of Significant Deterioration and Title V Greenhouse Gas Tailoring Rule, 75 Fed. Reg. 31,514 (June 3, 2010). Because current low emission thresholds in the act would trigger many smaller sources to become subject to the permitting programs for the first time, EPA issued the tailoring rule to raise those thresholds. In 2012, the D.C. Circuit upheld the tailoring rule. Coalition for Responsible Regulation, Inc. v. EPA, 684 F.3d 102 (D.C. Cir. 2012). The Supreme Court granted certiorari, but limited review to one narrow question: "[w]hether EPA permissibly determined that its regulation of greenhouse gas emissions from new motor vehicles triggered permitting requirements under the [CAA] for stationary sources that emit greenhouse gases."

[196] *UARG v. EPA*, 134 S. Ct. at 2444-46.

[197] *Id.* at 2444-49.

[198] CAA Section 111(b) authorizes EPA to set "new source performance standards," emission standards for new stationary sources. CAA Section 111(d) authorizes the agency to set emission standards for existing stationary sources that would be covered had they been new, among other preconditions.

In response to the *UARG* decision, EPA proposed revisions to existing PSD and title V regulations to ensure that neither the PSD nor title V rules require a source to obtain a permit solely because the source emits or has the potential to emit GHGs above the applicable thresholds.[199] This proposal also set a significant emissions rate for GHGs under the PSD program that would establish a *de minimis* threshold level of 75,000 tpy CO2-e (the same as the threshold in the Tailoring Rule) below which BACT is not required for a source's GHG emissions.[200]

It is unclear whether the Trump Administration will finalize the rule. The comment period on the proposed rule closed on December 16, 2016.[201] In the interim, EPA will continue applying 75,000 tpy CO2-e threshold that was originally finalized in the Tailoring Rule to determine if BACT is required for "anyway" sources.[202]

Considering Climate Change in Agency Determinations/Actions

In addition to reducing GHG emissions, various environmental statutes and regulations require federal agencies to consider climate change in their actions and decisions. Lawsuits have challenged the extent to which agencies consider climate change impacts and rely on predictive models, studies, and assumptions. This section will examine these issues in the context of the Endangered Species Act and the National Environmental Policy Act.

---

199 Revisions to the Prevention of Significant Deterioration (PSD) and Title V Greenhouse Gas (GHG) Permitting Regulations and Establishment of a Significant Emissions Rate (SER) for GHG Emissions Under the PSD Program, 81 Fed. Reg. 68,110 (Oct. 3, 2016).

200 81 Fed. Reg. at 68,113.

201 Revisions to the Prevention of Significant Deterioration (PSD) and Title V Greenhouse Gas (GHG) Permitting Regulations and Establishment of a Significant Emissions Rate (SER) for GHG Emissions Under the PSD Program; Proposed rule; extension of comment period, 81 Fed. Reg. 81,711 (Nov. 18, 2016).

202 Memorandum from J. McCabe & C. Giles to EPA Regional Administrators (July 24, 2014), *available at* https://www.epa.gov/sites/production/files/2015-07/documents/2014scotus.pdf.

Endangered Species Act (ESA)

The ESA provides legal protection to identified species and their critical habitat.[203] It is administered primarily by the Fish and Wildlife Service (FWS) for terrestrial and freshwater species,[204] but also by the National Marine Fisheries Service (NMFS) for certain marine species.[205] Under the ESA, FWS and NFMS can list individual species of plants and animals as either "endangered" or "threatened" based on risk assessments of their extinction.[206]

Lawsuits connecting the ESA to climate change typically are based on how an agency considered climate change when making determinations related to listing a species; designating critical habitat; or issuing a biological opinion on a federal agency action.[207] The ESA requires that the agency consider the effects on habitat, at least in part, for all of those determinations.[208] Accordingly, climate change evaluations long have been part of ESA decision making, but only to the extent that the climate's effects on habitat are linked to a species.

Federal agencies have relied on climate change projections to list a species as endangered or threatened, or to designate critical habitat under the ESA. When listing a species, the agency must make its decision "solely on the basis of the best scientific and commercial data available."[209] A similar standard applies when an agency designates a critical habitat under

---

[203] P.L. 93-205, 87 Stat. 884. 16 U.S.C. §§1531-1544. For more information on the ESA, see CRS Report RL31654, *The Endangered Species Act: A Primer*, by M. Lynne Corn and Alexandra M. Wyatt.

[204] For detailed information on the Fish and Wildlife Service (FWS) program for endangered species, see the FWS website at http://www.fws.gov/endangered/.

[205] For detailed information on the National Marine Fisheries Service (NMFS) program for endangered species, see the NMFS website at http://www.nmfs.noaa.gov/pr/laws/esa/. NMFS, a part of the National Oceanic and Atmospheric Administration (NOAA), is also sometimes referred to as *NOAA Fisheries*.

[206] 16 U.S.C. §1533(a).

[207] *See* 16 U.S.C. §1533(a)(1)(A) (when making determination on whether to list a species, relevant wildlife agency must consider "the present or threatened destruction, modification, or curtailment of its habitat or range"); 16 U.S.C. §1533(b)(2) (requiring relevant wildlife agency to designate critical habitat); and 16 U.S.C. §1536(a)(2) (requiring all agencies to consult with relevant wildlife agency to determine whether their actions would "result in the destruction or adverse modification of habitat of such species which is determined ... to be critical").

[208] *Id.*

[209] *Id.* §1533(b)(1)(A).

the ESA.[210] The agency designates critical habitat "on the basis of the best scientific data available" after taking into consideration the probable economic, national security, and other relevant impacts.[211]

Various groups have challenged the listing of species and the designation of critical habitat in the Arctic region, questioning whether model-based climate predictions constitute the "best scientific and commercial data available" on which to base ESA listing decisions. Most often, the suits relate to the listing of a species as threatened or endangered based on long-term climate models when its present population is stable. When challenged in court, an agency's ESA listing or and designation of its critical habit is often upheld because the court's review of the agency decision is narrow and highly deferential; the court will not set aside an agency ESA listing or designation decision so long as it is rational and reasonably based on supporting evidence.[212] However, courts have faulted agencies for inadequately considering climate change.[213]

Polar Bear Listing

The legal challenges to FWS's listing of the polar bear and designation of its critical habitat illustrate how courts have applied this narrow and deferential standard of review and interpreted the ESA standards for the best available data in the climate change context. In 2013, in *Safari Club International v. Salazar (In re Polar Bear Endangered Species Act Listing and Section 4(d) Rule Litigation)* (hereinafter *In re Polar Bear*), the D.C. Circuit upheld FWS's listing of polar bears as a threatened species under the ESA based in part on projected climate change impacts to the species and its habitat.[214] FWS's decision was based on three main conclusions: (1)

---

[210] *Id.* §1533(b)(2).

[211] *See id.*

[212] *See* Alaska Oil & Gas Ass'n v. Jewell, 815 F.3d 544, 554 (9th Cir. 2016).

[213] *See infra* "Considering Climate Change in Agency Determinations/Actions."

[214] 709 F.3d 1 (D.C. Cir. 2013). Note that the lower court had upheld FWS's listing against an array of consolidated legal challenges: from environmental and animal welfare groups and other plaintiffs that had argued that the polar bear should have been listed as "endangered" rather than "threatened," as well as from industry groups, sportsmen's groups, the State of Alaska, and other plaintiffs that had argued that the polar bear listing was unwarranted. *See In re Polar Bear Endangered Species Act Listing and §4(d) Rule Litigation)*, 838 F. Supp. 2d 214, 218 (D.D.C. 2011). The lower court had, however, ordered FWS to conduct

that the polar bear is dependent on sea ice for its survival; (2) that sea ice is declining; and (3) that climate change will continue to reduce the extent and quality of arctic sea ice gravely enough to endanger the polar bear population.[215]

The court held that the challenges to FWS's scientific assessment and conclusions "'amount to nothing more than competing views about policy and science,' on which we defer to the agency."[216] The court also rejected arguments that climate science was too uncertain to support listing the polar bear as a species that is likely to become endangered in the "foreseeable future."[217] The court concluded that FWS's reliance on climate projections was "justifiable[,] clearly articulated [,] ... sufficient to support their definition of foreseeability."[218] The Supreme Court declined to review the case.[219]

Polar Bear Critical Habitat Designation

Once it listed the polar bear as a threatened species, FWS moved forwarded with designating its critical habitat.[220] The ESA defines critical habitat as the specific areas "within the geographical area occupied by the species" that the species needs for recovery and that therefore should be protected.[221] Critical habitat includes areas containing the physical and biological features (1) essential for the species' success, such as space for growth and normal behavior, food, breeding sites, and habitats protected

---

additional environmental review of the ESA Section 4(d) special rule for polar bears, *Id.* at 238, which FWS did in 2013. *See* 78 Fed. Reg. 11,766 (Feb. 20, 2013). The appellants in the D.C. Circuit appeal in *In re Polar Bear* were among those that had argued that the polar bear listing was unwarranted.

215 *Id.* at 8.

216 *Id.* at 9 (quoting *In re Polar Bear Endangered Species Act Listing and §4(d) Rule Litigation*, 794 F. Supp. 2d 65, 69 (D.D.C. 2011)).

217 *Id.* at 15, quoting 16 U.S.C. §1532(20).

218 *Id.* at 16.

219 The Supreme Court denied the petition for certiorari. *Safari Club Int'l v. Jewell*, 2013 U.S. LEXIS 7036 (2013).

220 16 U.S.C. §1533(a)(3)(A)(i). For additional information on critical habitat designations, see CRS Report RL31654, *The Endangered Species Act: A Primer*, by M. Lynne Corn and Alexandra M. Wyatt.

221 16 U.S.C. §1532(5)(A)(i).

from disturbance, and (2) which may require special management or protection.[222]

In a 2016 decision, *Alaska Oil & Gas Association v. Jewell*, the U.S. Court of Appeals for the Ninth Circuit (Ninth Circuit) upheld FWS's designation of 187,000 square miles as critical habitat for the polar bear.[223] FWS's critical habitat designation was based in part on long-term projections of habitat destruction from climate change.[224] The FWS designated three areas on Alaska's coast and in its waters that contain "primary constituent elements" (PCEs) essential to the polar bear: a sea ice habitat, a terrestrial denning habitat, and a barrier island habitat.[225] The court, in large part, grounded its decision to uphold the FWS's habitat designation on the D.C. Circuit decision in *In re Polar Bear*.[226]

The Ninth Circuit disagreed with the lower court's narrow interpretation of the ESA critical habitat requirements. The court rejected the lower court's finding that the ESA required FWS to limit the critical habitat designation to specific areas that are currently used by polar bears, explaining that "[n]o such limitation to existing use appears in the ESA, and such a narrow construction of critical habitat runs directly counter to the Act's conservation purposes. The Act is concerned with protecting the future of the species, not merely the preservation of existing bears. And it requires use of the best available technology, not perfection."[227] The court concluded that the FWS properly relied on climate science and sea ice data

---

[222] *Id.;* 50 C.F.R. §424.12(b).

[223] Alaska Oil & Gas Ass'n v. Jewell, 815 F.3d 544, 558-59 (9th Cir. 2016).

[224] *Id.* at 558-59.

[225] *Id.* at 555.

[226] *See id.* at 558-59 (citing *In re Polar Bear*, 709 F.3d at 3).

[227] *See id.* at 555-56 (concluding that the "ESA thus requires FWS, when designating critical habitat, to focus on the PCEs essential to protecting the polar bear.... Since the point of the ESA is to ensure the species' recovery, it makes little sense to limit its protections to the habitat that the existing, threatened population currently uses.... The Act contemplates the inclusion of areas that contain PCEs essential for occupation by the polar bear, even if there is no available evidence documenting current activity.... [T]he agency must look beyond evidence of actual presence to where the species is likely to be found.... The standard FWS followed, looking to areas that contained the constituent elements required for sustained preservation of polar bears, was in accordance with statutory purpose and hence could not have been arbitrary, capricious, or contrary to law.") (internal citations omitted).

in designating habitat that may become critical in the future because of climate change and other factors.[228]

The State of Alaska, oil and gas groups, regional organizations, and Alaska Native groups are seeking review by the Supreme Court, arguing that the Ninth Circuit's "lax" and "exceptionally overbroad" standards allow FWS to draw impermissibly expansive boundaries in designating critical habitat.[229] Conservation groups have opposed the petition, arguing that the Ninth Circuit applied the appropriate statutory standards and that the facts support the need for a large area to be designated as critical for the polar bears.[230] The Court has not issued a decision on whether it will review the case.

Other Ninth Circuit ESA Cases

The Ninth Circuit had another opportunity in 2016 to review the use of climate change projections in ESA listing decisions. On October 24, 2016, the Ninth Circuit upheld ESA protections for two populations of Arctic bearded seals in *Alaska Oil & Gas Association v. Pritzker*.[231] NMFS listed the seals as threatened in 2012 based on climate change models that predicted that sea ice the seals depend on for birthing and mating would mostly disappear by 2095.[232] Plaintiffs argued that the models used in the listing decision could not reliably predict climate change effects on the seals beyond 2050.[233] A district court ruling agreed and vacated the rule; NMFS and the Center for Biological Diversity then appealed the decision.[234]

In upholding the bearded seal listing, the Ninth Circuit concluded that NMFS may base its listing decision on climate change models and long-term projections, provided that the record includes a reasonable

---

[228] *See id.* at 558-59 (listing the studies, reports, and climate models that predict sea ice loss and rescission as a result of climate change).

[229] Petition for Writ of Certiorari, Alaska Oil & Gas Ass'n, et. al., Alaska Oil & Gas Ass'n, Nos. 16-596, 16-610 (U.S. Nov. 4, 2016).

[230] Brief in Opposition to Petition for Writ of Certiorari, Ctr. for Biological Diversity, Alaska Oil & Gas Ass'n, No. 16 596 (Jan. 6, 2017).

[231] Alaska Oil & Gas Ass'n v. Pritzker, 840 F.3d 671 (9th Cir.2016).

[232] *Id.* at 681-82.

[233] *Id.* at 681.

[234] *Id.* at 674-75.

explanation for its decision and acknowledges any limitations.[235] Similar to its reasoning set forth in *Alaska Oil & Gas Association v. Jewell*, the court explained that the

> ESA does not require NMFS to base its decision on ironclad evidence when it determines that a species is likely to become endangered in the foreseeable future; it simply requires the agency to consider the best and most reliable scientific and commercial data and to identify the limits of that data when making a listing determination.[236]

The U.S. District Court of Alaska, however, appears to question listing decisions based on longterm climate projections. In the past two years, the Alaska district court rejected the listing determination for the Beringia seal discussed above and, most recently, the ringed seal, finding in both cases that the FWS's climate change predictions were too speculative to support a listing.[237] The district court was not able to consider the Ninth Circuit's bearded seal decision because *Alaska Oil & Gas Association v. Jewell* was issued after the lower court issued its decision on the ringed seal.[238] The FWS and environmental groups have appealed the ringed seal decision to the Ninth Circuit.[239] The environmental groups are asking the Ninth Circuit to reverse the lower court's holding for the ringed seal based on the reasoning applied in the bearded seal appellate decision.[240]

Other Recent ESA Cases Involving Climate Change Projections

Other courts have concluded that climate change projections are sufficient to satisfy the statutory requirement that listings and biological opinions under the ESA be based on the best available data. Two federal districts courts have faulted federal agencies for failing to credit and

---

[235] *Id.* at 681.
[236] *Id.*
[237] Alaska Oil & Gas Ass'n v. NMFS, No. 4:14-CV-00029-RRB, 2016 U.S. Dist. LEXIS 32002 (D. Alaska Mar. 17, 2016).
[238] *Id.* at *34.
[239] Alaska Oil & Gas Ass'n v. Pritzker, No. 16-35382 (9th Cir. docketed May 5, 2016).
[240] Opening Brief for Intervenor-Defendants, Alaska Oil & Gas Ass'n v. Pritzker, No. 16-35382 (9th Cir. Nov. 2, 2016).

account for the best data available on climate change impacts in their ESA decisions.

In *Defenders of Wildlife v. Jewell*, a federal district court in Montana held that the FWS's decision to withdraw the proposed listing of the wolverine as threatened was arbitrary and capricious.[241] In reversing its position on the proposed listing, the FWS attempted to discredit certain scientific studies on climate change effects that it had relied on previously to propose listing the wolverine as threatened, claiming that the FWS needed greater certainty and refinement in the climate change data before listing the wolverine.[242] The court concluded that the FWS "cannot demand a greater level of scientific certainty than has been achieved in the field to date—the 'best scientific data available' ... standard does not require that the [FWS] act only when it can justify its decision with absolute confidence,' and 'the ESA accepts agency decisions in the face of uncertainty.'"[243] The court vacated the withdrawal of the proposed listing and remanded the proposal to the FWS for further consideration.

In another ESA case, NMFS issued a biological opinion that concluded that changes to the operation of the Leavenworth National Fish Hatchery were not likely to jeopardize the existence of critical habitat for Upper Columbia River spring Chinook salmon or steelhead.[244] Section 7 of the ESA requires federal agencies to "insure that any action authorized, funded, or carried out by such agency ... is not likely to jeopardize the continued existence of any endangered species or threatened species or result in the destruction or adverse modification of [the critical] habitat of such species."[245] Any federal agency planning any action that may affect ESA-listed species must consult with NMFS or FWS.[246] After the consultation, the consulting agency must issue a biological opinion based

---

[241] Defs. of Wildlife v. Jewell, 176 F.Supp.3d 975, 1011 (D. Montana, 2016). The North American wolverine, which relies on deep snow to den and reproduce, is expected to be affected over the coming decades by warming temperatures and declines in snow levels predicted by climate models. *Id.* at 978-80.

[242] *Id.* at 1002-03.

[243] *Id.* at 1003.

[244] Wild Fish Conservancy v. Irving, No. 2:14-CV-0306-SMJ, 2016 U.S. Dist. LEXIS 162056, *9 (E.D. Wash. Nov. 22, 2016).

[245] 16 U.S.C. §1536(a)(2).

[246] *Id.* §1536(a)(2); 50 C.F.R. §402.14(a).

on "the best scientific and commercial data available" that determines whether the proposed action is likely to jeopardize the ESA-listed species or adversely modify critical habitat.[247]

Environmental groups challenged NMFS's biological opinion on several grounds, and in November 2016, a federal district court in Washington ruled that NMFS's biological opinion was arbitrary and capricious because NMFS failed to consider adequately climate change effects in the opinion's analysis of hatchery's modified operations and water use.[248] The court explained that "[t]he best available science indicates that climate change will affect stream flow and water conditions throughout the Northwest" and that the lack of a model or study specifically addressing local climate change effects did not permit the NMFS to ignore this factor.[249] The court found that NMFS had included "no discussion whatsoever" of the potential effects of climate change on the hatchery's future operations and water use, and that it was not sufficient for NMFS to say that the local area at issue was less prone to climate change effects than other areas in the region.[250] The judge remanded the matter for further consultation.[251]

Potential Implications

It is uncertain whether these decisions upholding listing and critical habitat decisions will affect decisions related to species outside of the Artic region. While relying on climate change projections may lead to the listing of some species with stable populations, for species outside of the Arctic region, there may be fewer studies and models that can reliably predict impacts of climate change on their habitats and their potential to adapt to justify an ESA listing.

The extent to which debate over climate change may determine the ESA's application may be limited by judicial decisions that require agencies to consider climate change in their ESA analyses, such as the

---

[247] 50 C.F.R. §402.14(g)(8), (h)(2)–(3).
[248] *Wild Fish Conservancy*, at *17-*20.
[249] *Id.* at *18.
[250] *Id.* at *19.
[251] *Id.* at *29.

biological opinions discussed above. Recently, on January 17, 2017, 13 state attorneys general sent a letter to the President Trump transition team requesting repeal of the critical habitat rules.[252] The March 28, 2017, executive order did not specifically address the ESA. However, the agencies may identify these regulations as burdening the "development or use" of energy resources and recommend to the President that they should be revised or rescinded.[253] The agency would then have to initiate a rulemaking to do so.

Furthermore, the series of Ninth Circuit cases may prompt some Members of Congress to seek to amend the ESA. On February 15, 2017, the Senate Committee on Environment and Public Works held a hearing entitled "Oversight: Modernization of the Endangered Species Act" that explored the possibility of amending the ESA to raise the threshold standard to list a species under the ESA.[254]

National Environmental Policy Act

The National Environmental Policy Act (NEPA)[255] requires federal agencies to consider, document, and disclose the potential effects of their actions and decisions on the environment.[256] In general, courts have held that federal agencies are required to consider climate change impacts in their NEPA analyses where these impacts are sufficiently serious and causally connected to the project.[257] While most federal agencies already have procedures in place to ensure that they identify GHG emissions

---

[252] The letter to the President Trump transition team was signed by the attorneys general from Alabama, Alaska, Arizona, Arkansas, Kansas, Louisiana, Michigan, Montana, Nebraska, Nevada, South Carolina, Texas, West Virginia, and Wyoming. Letter from Luther Strange, Ala. State Att'y Gen., to Ado Machida, Pol'y Implementation Team Lead (Jan. 17, 2017), http://ago.alabama.gov/news/972.pdf.

[253] Exec. Order, Promoting Energy Independence and Economic Growth §2 (Mar. 28, 2017).

[254] Juan Carlos Rodriguez, *Senate Gets Ball Rolling to Amend the ESA*, LAW360, Feb. 15, 2017, https://www.law360.com/articles/891979/senate-gets-ball-rolling-to-amend-the-esa.

[255] 42 U.S.C. §§4321-4370f.

[256] Id. §4332(2)(C). For additional information on NEPA, see CRS Report RL33152, *The National Environmental Policy Act (NEPA): Background and Implementation*, by Linda Luther.

[257] *See, e.g.*, Ctr. for Biological Diversity v. NHTSA, 538 F.3d 1172, 1218 (9th Cir. 2008) ("The impact of greenhouse gas emissions is precisely the kind of cumulative impacts analysis that NEPA requires agencies to conduct.").

associated with their proposed actions, as well as future impacts to those proposed projects that may relate to climate change, agencies have lacked clear direction on when and how to consider climate change impacts. This lack of guidance resulted in inconsistent environmental reviews across federal agencies, prompting stakeholders to file lawsuits challenging whether the agency NEPA reviews adequately considered potential effects from GHG emissions and climate change from their proposed actions.[258]

In order to provide guidance and improve consistency, the White House Council on Environmental Quality (CEQ) issued in August, 2016, its final guidance for federal agencies to assess the impact of their decisions on GHG emissions and also how such decisions may be impacted by a changing climate when conducting NEPA reviews.[259] In the Guidance, CEQ states that "when addressing climate change agencies should consider: (1) The potential effects of a proposed action on climate change as indicated by assessing [direct and indirect] GHG emissions...; and, (2) The effects of climate change on a proposed action and its environmental impacts."[260] The Guidance also recommends that "agencies review their NEPA procedures and propose any updates they deem

---

[258] *See, e.g.*, Mayo Found'n v. Surface Transp. Bd., 472 E3d 545 (8th Cir. 2006) (challenging the environmental impact statement for the approval construction of a 280-mile rail line from South Dakota to the Wyoming Powder River Basin for failing to full analyze adverse effects of increased coal consumption on the environment); Border Power Plant Working Grp. v. DOE, 260 F. Supp. 997 (S.D. Cal. 2003) (challenging the agencies' NEPA review for the construction of transmission lines to connect new power plants in Mexico to the California power grid as inadequate, in part because the environmental assessments failed to evaluate the impact of carbon emissions from the Mexican power plants); Friends of the Earth v. Mosbacher, 488 E Supp. 2d 889 (N.D. Cal. 2007) (seeking to compel the Overseas Private Investment Corporation and the Export-Import Bank of the United States, two quasi-governmental agencies to conduct environmental assessments to address the impacts to global warming from projects supported by the agencies outside the United States).

[259] Final Guidance for Federal Departments and Agencies on Consideration of Greenhouse Gas Emissions and the Effects of Climate Change in National Environmental Policy Act Reviews, 81 Fed. Reg. 51, 866 (Aug. 5, 2016). The CEQ guidance is available at https://obamawhitehouse.archives.gov/sites/whitehouse.gov/files/documents/nepa_final_ghg_guidance.pdf. For more information on the contents of the Guidance, see CRS Insight IN10554, *Overview of CEQ Guidance on Greenhouse Gases and Climate Change*, by Linda Luther.

[260] *Id.* at 5.

necessary or appropriate to facilitate their consideration of GHG emissions and climate change."[261]

After the CEQ GHG Guidance was released, agencies began working to update and align their NEPA procedures with the CEQ recommendations. For example, in February 2017, the Federal Energy Regulatory Commission (FERC) issued two volumes of guidance for how its officials should consider direct GHG and climate change impacts in NEPA reviews of proposed gas infrastructure projects.[262] The court may consider the new guidance in pending NEPA GHG litigation over FERC's approval of a proposed $683 million natural gas pipeline.[263] To date, the court has not issued a decision in the case.

For natural gas exports, the D.C. Circuit determined that DOE, and not FERC, is responsible for considering the broader upstream and downstream climate impacts related to allegedly higher domestic natural gas prices and increased use of coal in its NEPA review in approving the gas export.[264] The court explained that FERC, in approving a terminal's construction, is not responsible for a NEPA environmental review concerning the actual export of the gas. Environmental groups are currently challenging DOE's NEPA review for its gas export approvals in the D.C. Circuit.[265]

President Trump's March 28, 2017, executive order rescinds the CEQ's NEPA GHG guidance. Federal agencies, however, appear to still be required to consider the GHG and climate change impacts of their proposed actions.[266] Even prior to the issuance of CEQ's GHG guidance, courts had faulted federal agencies for insufficiently taking into account

[261] *Id.* at 3.

[262] *See* Notice of Availability of the Final Guidance Manual for Environmental Report Preparation, 82 Fed. Reg. 12,088 (Feb. 28, 2017). The guidance is available at https://www.ferc.gov/industries/gas/enviro/guidelines.asp.

[263] *See, e.g.*, Catskill Mountainkeeper Inc. et al. v. FERC, No.16-345 (2d Cir. filed Feb. 5, 2016); Stop the Pipeline v. FERC, No. 16-361 (2d Cir. filed Feb. 5, 2016) (challenging FERC's approval of a proposed natural gas pipeline, arguing that FERC failed to adequately considered the project's indirect impacts on increased gas drilling and its cumulative effects on GHG emissions and climate change).

[264] *See* Sierra Club v. FERC, 827 F.3d 59 (D.C. Cir. 2016) (holding that the decision to allow liquid natural gas to be exported rests exclusively with DOE).

[265] Sierra Club v. DOE, No. 16-1252 (D.C. Cir. filed July 26, 2016).

[266] Exec. Order, Promoting Energy Independence and Economic Growth §3(c) (Mar. 28, 2017).

climate change in NEPA reviews. For example, in *Center for Biological Diversity v. National Highway Traffic Safety Administration*, the U.S. Court of Appeals for the Ninth Circuit held that "the impact of greenhouse gas emissions on climate change is precisely the kind of cumulative impacts analysis that NEPA requires agencies to conduct," and remanded an agency action for further NEPA analysis.[267]

Social Cost of Carbon

The Social Cost of Carbon (SCC) is a monetary estimate of economic damages associated with an incremental increase in carbon dioxide (CO2) emissions.[268] The net present value of these damages is calculated by multiplying future costs by an appropriate discount factor and summing across all affected years.[269] The estimate is intended to include (but is not limited to) changes in net agricultural productivity, human health, property damages from increased flood risk, and the value of ecosystem services due to climate change.[270] Federal agencies have used the SCC in numerous final rulemakings and environmental reviews.[271]

There are a few court decisions that address the specific use of the SCC in federal agency rulemakings and environmental reviews. In general, courts have focused on an agency's justification (or lack thereof) regarding the use of the SCC values in its rulemaking or environmental review. Most recently, in August 2016, a federal court for the first time in *Zero Zone, Inc. v. Department of Energy (DOE)* upheld the use of the (SCC) in a cost-benefit analysis by a federal agency with regard to energy efficiency standards.[272]

---

[267] Ctr. for Biological Diversity v. NHTSA, 538 F.3d 1172, 1218, 550 (9th Cir. 2008).

[268] INTERAGENCY WORKING GROUP ON SOCIAL COST OF GREENHOUSE GASES, TECHNICAL SUPPORT DOCUMENT: TECHNICAL UPDATE OF THE SOCIAL COST OF CARBON FOR REGULATORY IMPACT ANALYSIS UNDER EXECUTIVE ORDER 12866 (2016), https://www.epa.gov/sites/production/files/2016-12/documents/sc_co2_tsd_august_2016.pdf.

[269] *Id.* at 3.

[270] *Id.*

[271] See CRS Report R44657, *Federal Citations to the Social Cost of Greenhouse Gases*, by Jane A. Leggett. Agencies began using the social cost of methane (SC-CH4) in 2015, and the social cost of nitrous oxide (SC-N2O) in 2016. Collectively, these are the social costs of greenhouse gases (SC-GHG). *Id.*

[272] Zero Zone, Inc. v. DOE, 832 F.3d 654, 664, 667 (7th Cir. 2016).

The Energy Policy and Conservation Act of 1975 (EPCA) directs DOE to develop mandatory energy conservation standards for specific types of equipment and appliances.[273] EPCA requires, among other things, that these standards must be designed to "increase[] the maximum improvement in energy efficiency" and be "technologically feasible and economically justified."[274] In 2014, DOE established new energy efficiency standards for 49 classes of commercial refrigeration equipment (CRE) and clarified the testing procedures that DOE uses to implement those standards.[275]

In *Zero Zone, Inc.,* petitioners claimed that DOE abused its discretion by considering environmental factors when determining whether these new energy efficiency standards were "economically justified" and, in the alternative, that the use of the SCC was arbitrary and capricious.[276] In rejecting the petitioners' claims, the U.S. Court of Appeals for the Seventh Circuit held that "the expected reduction of environmental costs needs to be taken into account" in DOE's cost-benefit analysis, and that it had "no doubt that Congress intended that DOE have the authority under the [EPCA] to consider the reduction in SCC."[277] In addition, the court concluded that the use of the SCC was not arbitrary and capricious because DOE had addressed concerns regarding the SCC in the final rule by acknowledging the limitation of the SCC, referencing multiple entities that support that SCC values, and clarifying that despite those concerns, DOE could use the SCC in its cost-benefit analysis.[278]

Previous judicial decisions have faulted agencies for failing to use the SCC or other types of tools to account for costs related to carbon emissions in their cost-benefits analysis. For example, in 2008, the Ninth Circuit addressed the National Highway Traffic Safety Administration's (NHTSA's) refusal to include the SCC in setting average fuel economy

---

[273] 42 U.S.C. §§6295(a); 6313.
[274] *Id.* §6295(o)(2).
[275] *Zero Zone, Inc.*, at 664, 667.
[276] *Id.* at 677.
[277] *Id.*
[278] *Id.*

standards for light trucks, model years 2008 through 2011.[279] In *Center for Biological Diversity v. NHTSA*, the Ninth Circuit held that it was arbitrary and capricious for an agency to fail to assess the impacts of GHG emissions in its cost-benefit analysis, even when there is uncertainty about those impacts, stating that "while the record shows there is a range of values [for the SCC], the value of carbon emissions reduction is certainly not zero."[280] The court concluded that the scientifically supported values presented by the petitioners show that "it is possible to monetize the benefit of carbon emission reduction."[281]

A few courts have also reviewed the use of the SCC in environmental reviews of proposed agency actions under the NEPA.[282] Agencies have used the SCC to evaluate potential GHG impacts associated with federal actions in order to comply with NEPA's environmental review requirements. A 2014 district court decision, *High Country Conservation Advocates v. U.S. Forest Service*, vacated the BLM's approval of a coal exploration plan for failure to justify why the SCC was not used to quantify the costs associated with the mining exploration.[283] The court held that it was arbitrary and capricious to quantify the *benefits* of the mining exploration but fail to quantify the *costs* when the SCC factor was available and used in the draft EIS; the agencies might have been able to offer "non-arbitrary reasons" why the protocol was not included in the

---

[279] Ctr. for Biological Diversity v. NHTSA, 538 F.3d 1172 (9th Cir. 2008).

[280] *Id.*

[281] *Id.* As further support for ruling against NHTSA, the court noted that (1) NHTSA gave no reasons why it believed the range of values in the studies was "extremely wide," given that several commentators on the proposed rule recommended the *same* value—$50 per ton of carbon; (2) NHTSA has monetized other uncertain benefits such as the reduction of other pollutants, crash, noise, and construction costs; (3) the evidence contradicted NHTSA's conclusion that commentators on the proposed rule did not reliably show that monetizing the value of carbon reduction would have affected the stringency of the CAFE standard; and (4) NHTSA failed to support its claim that had it accounted for the benefit of carbon reduction, it would have had to account for the adverse safety effects of manufacturers making lighter vehicles, and the two would have balanced out. *Id.* at 1200-03. As remedy for NHTSA's failure to monetize the value of carbon emissions in setting the CAFE standard, the court remanded the standard to NHTSA in order for it to promulgate a new CAFE standard as expeditiously as possible. *Id.* at 1203.

[282] *See supra* "National Environmental Policy Act" section.

[283] *High Country Conservation Advocates v. U.S. Forest Serv.*, 52 F. Supp. 2d 1174 (D. Colo. 2014).

final EIS, but did not.[284] And as in *Center for Biological Diversity*, above, the court held that it was unacceptable to treat the cost of carbon as zero when the SCC estimates, though covering a wide range, did not include the possibility that the cost was zero.[285]

BLM's environmental reviews continue to be challenged in court. On August 25, 2016, WildEarth Guardians and Physicians for Social Responsibility filed a complaint asking the U.S. District Court for the District of Columbia to vacate authorizations for 397 oil and gas leases on public lands in three states on the grounds that the BLM violated NEPA by failing to use the SCC or any other economic or scientific tools to assess the climate impacts of these leases.[286] The plaintiffs contend that BLM did not use the SCC protocol, or any other economic or scientific tools, to assess the potential impacts of challenged BLM lease authorizations on the climate.[287] The parties have not filed their briefs with the court yet.

In contrast, the D.C. Circuit upheld an agency's environmental review under NEPA that did not use the SCC to quantify potential GHG impacts from a project in *EarthReports, Inc. v. Federal Energy Regulatory Commission (FERC)*.[288] Unlike *High Country Conservation Advocates*, where BLM provided no explanation on why it did not use the SCC, FERC acknowledged the availability of the SCC tool but explained that it would not be "appropriate" or "informative" because of significant variation in output and lack of incremental impact measurement—criteria related to NEPA environmental reviews.[289] The court concluded that FERC acted reasonably in finding that the SCC was "inadequately accurate" to use in its environmental review of a liquid natural gas facility conversion project.[290]

These judicial opinions illustrate how these courts focused on an agency's justification (or lack thereof) to determine whether the agency

---

[284] *Id.* at 1191-92.

[285] *Id.*

[286] Compl. for Decl. J. & Inj. Relief at 1-2, WildEarth Guardians v. Jewell, No. 1:16-cv-01724 (D.D.C, filed Aug. 25, 2016).

[287] Id. at 32.

[288] EarthReports, Inc. v. FERC, 828 F.3d 949 (D.C. Cir. 2016).

[289] Id. at 956. (internal quotations omitted).

[290] *Id.*

acted reasonably in using or not using the SCC in its rulemaking or environmental review. Although each of the agencies was faced with the same uncertainties inherent in the SCC tool, the courts faulted those agencies that failed to articulate why the SCC was not an appropriate tool to quantify the potential impacts of GHG emissions associated with the agencies' actions.

The D.C. Circuit may have another opportunity to address the use of the SCC in the CPP litigation, if it moves forward.[291] One of the issues raised in the petitions for review of the CPP is whether EPA's regulatory impact analysis is flawed because it "assesses domestic costs against global benefits" and relies on the SCC.[292] In a letter to the court, EPA cites *Zero Zone, Inc. v. DOE*, arguing that similar to DOE in *Zero Zone, Inc.*, EPA reasonably explained why its monetized benefit-cost analysis accounted for global benefits.[293] In response, petitioners distinguish *Zero Zone, Inc.*, claiming that the EPCA specifically allows DOE to consider other relevant factors, including international factors, unlike Section 111 of the CAA, which they argue does not contain an analogous provision and does not authorize EPA to consider foreign benefits.[294]

Similar to the CEQ NEPA GHG guidance discussed above, the Trump Administration has withdrawn "as no longer representative of governmental policy" all SCC guidance and documents and disbanded the Interagency Workgroup on Social Cost of Greenhouse Gases that developed the SCC metric.[295] Nevertheless, as the cases discussed above illustrate, it appears that federal agencies may potentially still be required to take into account costs of carbon in their rulemakings and environmental

[291] *See supra* "Power Plants" section.

[292] Opening Br. of Pet'rs on Procedural and Record-Based Issues at 69-70, West Virginia, et al. v. EPA, No. 15-1363 (D.C. Cir. Feb. 19, 2016); Reply Br. of of Pet'rs on Procedural and Record-Based Issues at 31-32, West Virginia, et al. v. EPA, No. 15-1363 (D.C. Cir. Apr. 15, 2016).

[293] Letter from Brian Link, Counsel of Record, Dept. of Justice, to Mark Langer, Clerk of the Court, D.C. Cir. (Aug. 17, 2016).

[294] Letter from Adam Gustafson, Counsel of Record, Boyden Gray & Assoc. PLLC, to Mark Langer, Clerk of the Court, D.C. Cir. (Aug. 25, 2016).

[295] Exec. Order, Promoting Energy Independence and Economic Growth §5 (Mar. 28, 2017).

reviews.[296] The executive order directs agencies to follow the guidance contained in OMB Circular A-4, dated September 17, 2003, when monetizing the value of changes in GHG emissions resulting from regulations.[297] Although the OMB Circular A-4 provides guidance on how to conduct cost-benefit analysis, it mentions climate change costs only once, with respect to addressing "uncertain knowledge of how some economic activities might affect future climate change."[298]

THRESHOLD BARRIERS TO LITIGATION

Federal courts often do not reach the merits of a climate change suit due to threshold procedural barriers, such as whether a plaintiff or petitioner has the right to bring a lawsuit in the first place or whether the court has jurisdiction over a type of claim.[299] To proceed to the merits of a case, a plaintiff must establish standing, a procedural threshold that has, at times, posed a difficult hurdle to overcome.

Standing Doctrine

The standing doctrine is derived from Article III of the Constitution, which limits the jurisdiction of federal courts to "cases" or "controversies."[300] To satisfy the constitutional standing requirements in

---

296 Doug Obey, *Advocates Warn Trump May Find It Challenging To Scrap Carbon 'Cost' Tool*, INSIDEEPA.COM, Nov. 28, 2016, https://insideepa.com/daily-news/advocates-warn-trump-may-find-it-challenging-scrap-carbon-cost-tool.

297 Exec. Order, Promoting Energy Independence and Economic Growth §5(c) (Mar. 28, 2017).

298 Off. of Management & Budget, Circular A-4, at 38 (Sept. 17, 2003).

299 As a threshold matter, a plaintiff challenging federal agency action in court must have a legal right to a judicial determination of the issues raised in its complaint. *See* Flast v. Cohen, 392 U.S. 83, 99 (1968). Whether a court has jurisdiction is often informed by constitutional, prudential, and statutory considerations. *See* Lujan v. Defs. Wildlife, 504 U.S. 555, 560 (1992); Bennett v. Spear, 520 U.S. 154, 162 (1997). A court may be called upon to determine, for example, whether the case presents a nonjusticiable political question. *See* Baker v. Carr, 369 U.S. 186, 217 (1962).

300 U.S. CONST. art iii, §2.

Article III, the Supreme Court imposes three requirements: (1) the plaintiff must allege a personal injury-in-fact, which is actual or imminent, concrete, and particularized; (2) injury must be "fairly traceable to the defendant's allegedly unlawful conduct; and (3) the injury must be "likely to be redressed by the requested relief."[301] A plaintiff or petitioner that fails to satisfy any of these elements lacks standing, and the court will dismiss the suit.

The issue of a state's standing to bring a climate change suit was central to two landmark climate change decisions, *Massachusetts v. EPA* in the Supreme Court[302] and *American Electric Power Co. v. Connecticut* (*AEP*) in the U.S. Court of Appeals for the Second Circuit (Second Circuit).[303]

Both decisions noted that Article III standing is satisfied when a state brings suit as *parens patriae* (literally, father of the country) on behalf of its citizens.[304] *Parens patriae* doctrine allows a state to sue in its sovereign capacity to protect its citizenry, rather than being limited, as Article III would normally require, to asserting traditional particularized injuries to state interests.[305]

---

[301] *See, e.g.*, Lujan v. Defenders of Wildlife, 504 U.S. 555, 560 (1992). When it is an organization that sues on behalf of its members, rather than an individual, the standing requirements are (1) the members (or some of them) must have standing to sue in their own right; (2) the interests the organization seeks to protect in the suit are germane to the organization's purpose; and (3) neither the claim asserted nor the relief requested requires the participation of individual members in the lawsuit. Hunt v. Washington State Apple Advertising Comm'n, 432 U.S. 333, 343 (1977).

[302] 549 U.S. 497 (2007).

[303] 582 F.3d 309 (2d Cir. 2009), *rev'd on other grounds*, 131 S. Ct. 2527 (2011) (affirming the Second Circuit's finding of standing by equally divided vote). *See generally* Kirsten Engle, *State Standing in Climate Change Lawsuits*, 26 J. LAND USE & ENVTL. L. 217 (2011).

[304] *Massachusetts*, 549 U.S. at 518-520; Connecticut v. American Elec. Power Co., 582 F.3d 309, 338-339 (2d Cir. 2009), *rev'd on other grounds,* 131 S. Ct. 2527 (2011). That is, Article III is satisfied when a state brings suit as *parens patriae* on behalf of its citizens. *Massachusetts*, 549 U.S. at 519-521.

[305] The modern origins of the *parens patriae* doctrine lie in two century-old nuisance cases brought by states in federal court alleging interstate pollution: *Missouri v. Illinois*, 180 U.S. 208 (1901), and *Georgia v. Tennessee Copper Co.*, 206 U.S. 230 (1907). In both cases, state standing was found. The current test for *parens patriae* standing is found in *Snapp & Son, Inc. v. Puerto Rico*, 458 U.S. 592 (1982), though there is some question whether traditional Article III standing requirements have to be met as well by the citizens of the state asserting *parens patriae* standing. *See generally*, Sara Zdeb, *From* Georgia v. Tennessee Copper *to*

Most of the Supreme Court's decision in *Massachusetts v. EPA* is devoted to whether plaintiffs had standing to sue.[306] At the outset, the Court asserted that states are entitled to "special solicitude" when seeking to establish standing.[307] The Court found that the state met all three factors to establish standing—injury-in-fact, causation, and redressability of the injury. For the first factor, the Court determined that the current and potential future loss of Massachusetts shore land from sea level rise was an injury-in-fact. That this injury may be widely shared with other coastal states does not disqualify this injury, said the Court; it is nonetheless particular and concrete.[308]

The second prong of the standing test is causation, requiring that the alleged injury be fairly traceable to the defendant.[309] EPA did not dispute the existence of a causal relationship between GHG emissions and climate change. It did argue, however, that any reduction in GHG emissions achieved through the current litigation would be too tiny a fraction of worldwide GHG emissions to make a cognizable difference in climate change.[310] In an important ruling that may benefit environmental plaintiffs in other contexts, the Court held that even an agency's refusal to take a "small incremental step" that would result in only a modest reduction in worldwide GHG emissions, is enough for standing purposes.[311]

The third and final factor is whether the remedy sought by the plaintiff is likely to redress the injury. In this case, Massachusetts sought as a

---

Massachusetts v. EPA; *Parens Patriae Standing for State Global Warming Plaintiffs*, 96 Geo. L. J. 1059 (2008).

[306] *Massachusetts v. EPA*, 549 U.S. at 515-29.

[307] 549 U.S. at 520. The petitioners had two factors in their favor. First, the CAA specifically authorizes challenges to agency action unlawfully withheld, such as in the *Massachusetts* suit. 42 U.S.C. §7607(b)(1). A litigant to whom Congress has accorded such a procedural right, said the Court, can assert that right without meeting the normal standards for standing. 549 U.S. at 517-518. Second, the Court found it "of considerable relevance" that the petitioner's injury—Massachusetts's loss of shore land from global-warming-induced sea level rise—was that of a sovereign state rather than a private entity. *Id.* at 518. States are "not normal litigants for the purposes of invoking federal jurisdiction," said the Court, noting their quasi-sovereign duty to preserve their territory. *Id.*

[308] *Id.* at 522.

[309] *Id.* at 517, 523.

[310] *Id.* at 523.

[311] *Id.* at 524.

remedy a regulation of GHG emissions from new motor vehicles.[312] The Court found that this remedy satisfied redressability because while it would not by itself reverse climate change, it would nonetheless slow or reduce it.[313]

After its *Massachusetts v. EPA* decision, the Supreme Court considered the question of the standing of state and non-state parties in climate change suits in *American Electric Power Co. v. Connecticut* (AEP).[314] As discussed above, the Second Circuit held that both state and non-state plaintiffs had standing in a tort suit seeking GHG reductions from five large electric utilities. The Court, by an equally divided vote of four-to-four, affirmed a Second Circuit standing decision but avoided directly addressing the standing of non-state parties in climate change suits.[315] The Court stated that "[f]our members of the Court would hold that at least some plaintiffs have Article III standing under *Massachusetts*, which permitted a State to challenge EPA's refusal to regulate [GHG] emissions ... and, further, that no other threshold obstacle bars review."[316] Without further explanation, the *AEP* decision leaves the question of whether non-state petitioners have standing under Article III in climate change suits unresolved.

A related issue to state standing is whether Indian tribes, by virtue of their inherent sovereignty, may establish standing through *parens patriae* status.[317] The argument for tribal *parens patriae* standing was rejected by the district court in *Native Village of Kivalina v. ExxonMobil Corp*. In this case, an Alaskan native village sought damages from energy companies and electric utilities for coastal erosion, alleging that GHG emissions from their operations contribute to climate change, and thus to the village's

---

312 *Id.* at 517, 525-26.

313 *See id.* at 525 (explaining that given the "enormity" of the potential effects of climate change, it was not relevant to the Court that the full effectiveness of the remedy would be delayed until existing cars and trucks on the road were largely replaced by new ones).

314 *American Electric Power Co. v. Connecticut,* 564 U.S. 410 (2011) [hereinafter *AEP*].

315 *Id.* at 420.

316 *Id.*

317 *See generally* Elizabeth Ann Kronk, *Effective Access to Justice: Applying the Parens Patriae Standing Doctrine to Climate Change-Related Claims Brought by Native Nations*, 32 PUB. LAND & RES. L. REV. 1 (2011).

erosion problem.[318] The court explained that unlike *Massachusetts v. EPA*, the village was not entitled to "special solitude" because they were seeking damages directed against private entities and not enforcing any procedural rights concerning an agency's rulemaking authority.[319] In dismissing the suit, the court held that the village lacked standing because it had not satisfied the "causation" component of standing doctrine—that is, the village had not connected their coastal erosion problem to the specific GHG emissions of the defendants.[320] In addition, the court concluded that the claim was nonjusticiable under the political question doctrine, under which lawsuits based on policy questions must be left to the political branches of government.[321]

Standing for Non-State Plaintiffs

In general, non-state plaintiffs have encountered difficulties in establishing standing in climate change-related litigation.[322] The finding of standing in *Massachusetts* generally did not extend to non-state plaintiffs seeking to establish standing in nonstatutory causes of action (such as common law nuisance), and the split decision in *AEP* failed to provide any further guidance.[323] Courts have largely rejected extending the "special solicitude" standard for standing set forth in *Massachusetts v. EPA* to any non-state.[324] Non-state plaintiffs have had some success in establishing

---

[318] Native Vill. of Kivalina v. ExxonMobil Corp., 663 F. Supp. 2d 863, 882 (N.D. Cal. 2009).

[319] *Id.*

[320] *Id.* at 877-80.

[321] *Id.* at 871-77.

[322] *See, e.g.*, Bradford C. Mank, *Article III Standing for Private Plaintiffs Challenging Greenhouse Gas Regulations,* 53 SAN DIEGO L. REV. 287 (2016) (analyzing standing in *Massachusetts*, *AEP*, and other cases).

[323] *See supra* "Common Law Nuisance and Political Question Doctrine" section.

[324] *See, e.g.*, Comer v. Murphy Oil USA, Inc., 839 F. Supp. 2d 849 (S.D. Miss. 2012) (finding of Article III standing for state sovereign in *Massachusetts v. EPA* does not support standing for private plaintiffs here), *aff'd on other grounds*, 718 F.3d 460 (5th Cir. 2013); Native Vill. of Kivalina v. ExxonMobil Corp., 663 F. Supp. 2d 863, 882 (N.D. Cal. 2009) (same), *aff'd on other grounds*, 696 F.3d 849 (9th Cir. 2012), *cert. denied*, 133 S. Ct. 2390 (2013); Washington Envtl. Council v. Bellon, 732 F.3d 1131, 1145 (9th Cir. 2013) (same).

standing in cases challenging an agency's failure to consider climate change impacts in their regulatory actions.[325]

Barriers to Common Law Claims

Other threshold barriers to litigation often involve whether federal courts have jurisdiction over the claim brought to the court. While most of the climate change litigation discussed above has focused on government regulation of GHG emissions, private plaintiffs also have tried to force emitting sources to reduce GHG emissions or to claim damages from emitters for their alleged contributions to climate change.[326] These efforts, however, have been largely unsuccessful.[327]

The few cases that have brought common law nuisance claims seeking injunction or damages have failed based on the court's lack of jurisdiction to hear the claims.[328] This section will review jurisdictional barriers involved in these common law nuisance suits.

Common Law Nuisance and Political Question Doctrine

Several common law nuisance claims have been dismissed at the trial court level because they raised nonjusticiable questions according to the political question doctrine. While standing asks whether there is a proper *plaintiff* before the court, the political question doctrine asks whether there is a justiciable *claim*. The doctrine is rooted in the separation of powers principles, and acts to restrain courts from interfering in policy questions

---

[325] *See, e.g.*, WildEarth Guardians v. Jewell, 738 F.3d 298 (D.C. Cir. 2013) (holding that harm to group's members from local pollution caused by federal leasing of coal lands was sufficient injury in fact to allow challenge to *all* of alleged deficiencies in environmental impact statement on proposed lease, including those related to climate change); Ctr. for Biological Diversity v. EPA, 90 F. Supp. 3d 1177, 1187-88 (W.D. Wash. 2015) (holding that members of the plaintiff environmental organization suffered concrete standing injuries from the EPA's approval of Washington's and Oregon's decisions not to identify any waters experiencing ocean acidification as impaired under the Clean Water Act).

[326] *See e.g.*, Comer v. Murphy Oil USA, 839 F. Supp. 2d 849, 862-865 (S.D. Miss. 2012), *aff'd* 718 F. 3d 460 (5th Cir. 2013).

[327] *See infra* "Common Law Nuisance and Political Question Doctrine" and "Displacement of Federal Common Law" sections.

[328] *See infra* "Displacement of Federal Common Law" section.

and decisions made by the political branches of government.[329] The Supreme Court in *Baker v. Carr* in 1962 held that some cases are nonjusticiable because they are constitutionally committed to other branches of the government, impossible to decide without an initial policy determination, or otherwise imprudent for judicial review.[330]

The doctrine has been the focal point in several nuisance-based climate change litigation matters, with courts rejecting such claims on that ground. For example, in *Comer v. Murphy Oil USA*, property owners sought money damages, claiming that GHG emissions from oil and energy companies were a "nuisance" that added to the severity of Hurricane Katrina, which damaged their property.[331] The court held, among other things that "the claims presented by the plaintiffs constitute nonjusticiable political questions, because there are no judicially discoverable and manageable standards for resolving the issues presented, and because the case would require the Court to make initial policy determinations that have been entrusted to the EPA by Congress."[332] Similarly, in *California v. General Motors Corp.*, the State of California brought a nuisance suit against automakers for money damages for their alleged contribution to climate change that was dismissed as a nonjusticiable political question.[333] in *Native Village of Kivalina v. ExxonMobil Corp.*, in addition to dismissing the case for lack of standing, the district court also held, among other things, that "the allocation of fault—and cost—of global warming is a matter appropriately left for determination by the executive or legislative

---

[329] JONATHAN MARTEL ET. AL.,GLOBAL CLIMATE CHANGE AND U.S. LAW 249-50 (Jody Freeman & Michael B Gerrard, eds., 2d ed. 2014).

[330] 369 U.S. 186, 216 (1962).

[331] Comer v. Murphy Oil USA, 839 F. Supp. 2d 849, 862-865 (S.D. Miss. 2012), *aff'd* 718 F. 3d 460 (5th Cir. 2013).

[332] *Id.* at 865.

[333] *See* California v. General Motors Corp., No. C06-05755-MJJ, 2007 U.S. Dist. LEXIS 68547 (N.D. Cal. Sept. 17, 2007) (holding that the "complaint raises non-justiciable political questions, it need not and does not reach the issue of whether the federal common law recognizes Plaintiffs global warming nuisance claim or whether congressional action has otherwise displaced Plaintiff's claim").

branch in the first instance."[334] Some of these cases were also dismissed because the claims were displaced by the CAA.[335]

Displacement of Federal Common Law

Federal common law may be displaced where "Congress addresses a question previously governed by a decision rested on federal common law."[336] The test of displacement is whether the federal statute speaks directly to the question raised in the lawsuit.[337] In 2011, the Supreme Court's decision in *AEP* (discussed above) barred federal common law claims (such as nuisance) against entities for alleged injuries from GHG emissions.[338] In a unanimous decision, the Court held that under Section 111 of the CAA, EPA's authority to regulate GHG emissions displaces federal common law rights to seek abatement of GHG emissions from power plants.[339]

Though the 2011 *AEP* ruling involved plaintiffs seeking injunctive relief (requiring the GHGemitting defendants to cap and reduce their CO2 emissions over time), a court has held the ruling to mean the CAA displaces federal common law actions seeking monetary damages as well.[340] The availability of *state* common law claims for reducing GHG emissions remains an open question although one court has held that state law nuisance claims may also be displaced.

If Congress amends the CAA to remove EPA's authority to regulate GHG emissions under the CAA, which would effectively overturn the Supreme Court's decisions in *AEP* and *Massachusetts v. EPA*, it would appear that common law nuisance claims would possibly no longer be displaced by federal actions, which would potentially open the door to a resurgence of these claims. Currently, the House is considering H.R. 637,

---

[334] Native Vill. of Kivalina v. Exxon Mobil Corp., 663 F. Supp. 2d 863, 877 (N.D. Cal. 2009), *aff'd by* 696 F.3d 849 (9th Cir. 2012). See supra "Standing Doctrine" section discussing the standing issues in this case.

[335] *See infra* "Displacement of Federal Common Law" section.

[336] Am. Elec. Power Co. v. Connecticut, 564 U.S. 410, 423 (2011).

[337] *Id.* at 424.

[338] *Id.*

[339] *Id.*

[340] Comer v. Murphy Oil USA, 839 F. Supp. 2d 849 (S.D. Miss. 2012), *aff'd* 718 F. 3d 460 (5th Cir. 2013).

the Stopping EPA Overreach Act of 2017, which would exclude GHGs from regulation under the Clean Air Act.[341]

Barring any congressional action removing EPA's authority, the *AEP* decision will continue to displace common-law-based climate change claims, making it unnecessary for courts to reach the standing and political question issues in such cases.[342] However, a recent case in the context of the public trust doctrine is raising those jurisdictional issues anew.

The Public Trust Doctrine

In November 2016, an Oregon federal district court judge denied the U.S. government's motion to dismiss a lawsuit brought by a group of 21 individuals, all age 20 or younger, and other plaintiffs seeking to compel the federal government to reduce CO2 emissions.[343] The case, *Juliana v. United States*, is part of a campaign of state and federal lawsuits and rulemaking petitions related to climate change coordinated by an Oregon nonprofit, Our Children's Trust.[344] While only a preliminary ruling, the opinion has prompted some media outlets to speculate whether the "kids'" lawsuit may potentially force actions to address climate change further through litigation.[345]

The plaintiffs allege that federal officials promoted policies that contribute to climate change, while knowing of the alleged dangers of those policies.[346] The plaintiffs ask the court to order the federal government to "implement an enforceable national remedial plan to phase out fossil fuel emission" and "stabilize the climate system," among other

---

341 H.R. 637, 115th Cong. (2017).

342 *See e.g.*, Native Vill. of Kivalina v. ExxonMobil Corp., 696 F.3d 849 (9th Cir. 2012) (rejecting the federal common law nuisance claim solely on the basis of the CAA displacement argument announced by the Supreme Court in *AEP*).

343 Juliana v. United States, No. 6:15-cv-01517-TC, 2016 U.S. Dist. LEXIS 156014 (D. Or. Nov. 10, 2016) [hereinafter *Juliana*]. For more information, see CRS Legal Sidebar WSLG1715, *UPDATE: Climate Change Litigation Update: "Children's Crusade" Case Against the United States Goes Forward*, by Stephen P. Mulligan.

344 For further information about Our Children's Trust, see https://www.ourchildrenstrust.org/.

345 http://www.cnn.com/2016/11/10/opinions/sutter-trump-climate-kids/; https://www.washingtonpost.com/news/energy-environment/wp/2016/12/01/trump-could-face-the-biggest-trial-of-the-century-over-climate-change/?utm_term=.7049b9b37913; http://www.businessinsider.com/kids-suing-government-for-climate-change-2016-11.

346 *Juliana*, at *7-*8.

requests for relief.[347] There are multiple causes of action in the complaint, which the district court organized into two categories: (i) alleged violations of plaintiffs' substantive due process rights to life, and liberty, and property in the Constitution and (ii) common law violations of the public trust doctrine.[348] The district court did not rule on the merits of these claims, nor did it issue a finding that the government bears legal or factual responsibility for increased carbon dioxide emissions. But the court did conclude that it has subject matter jurisdiction over the case and that the plaintiffs' alleged facts that, if proven to be true, could entitle them to relief.[349]

In addition to raising standing and political question issues similar to those discussed above,[350] the court's treatment of the plaintiffs' public trust claim may also represent a unique ruling in the context of climate change litigation. The public trust doctrine is a common law principle with ancient origins whereby the government has a duty to safeguard certain natural resources for the benefit of the public.[351] The public trust doctrine is traditionally associated with tidelands and the beds of navigable waterways, and the government argued that the doctrine should not be extended to create a duty to preserve the earth's atmosphere.[352] While the *Juliana* court stopped short of deciding that the atmosphere is part of the public trust, instead focusing on alleged rising sea levels and harm to the territorial sea, it left open the possibility of a future ruling that the public trust doctrine creates a federal government duty to protect the atmosphere.[353] Thus far, no court has ruled on the merits of an "atmospheric public trust" claim against the federal government, and state

---

[347] *Id.* at *42-*44.
[348] *Id.* at *46-*58.
[349] *Id.* at *87.
[350] *Id.* at *11-*58.
[351] See *id.* at *58-*68 (discussing the origins of the public trust doctrine).
[352] *Id.*
[353] *Id.*

courts have taken divergent views on the viability such a claim against state governments under their respective state constitutions and laws.[354]

The litigation is now moving forward starting with discovery over the coming months. On January 31, 2017, the court denied the plaintiffs' efforts to depose Secretary of State Rex Tillerson regarding his work related to climate change while at ExxonMobil and the American Petroleum Institute, among other issues.[355] The Trump Administration is seeking a late appeal of the denial of the motion to dismiss and to delay trial with the Ninth Circuit.[356] Industry groups, including the American Fuel & Petrochemical Manufacturers and the American Petroleum Institute, joined the Trump Administration and filed a similar motion for appeal on March 10, 2017.[357]

CONCLUSION

Congress faces a number of questions regarding climate change regulation, policy, and litigation. Some early actions by the Administration indicate it may seek additional time during the litigation process to consider how best to move forward in court.[358] Agency actions resulting from compliance with the March 28, 2017, executive order, including delaying the effective date of rules or withdrawing previous determinations, will directly affect ongoing litigation.[359] Resolution of the

---

[354] For more information related state claims and other similar suits, see CRS Legal Sidebar WSLG1715, *UPDATE: Climate Change Litigation Update: "Children's Crusade" Case Against the United States Goes Forward*, by Stephen P. Mulligan.

[355] Keith Goldberg, *Tillerson Won't Have To Testify In Kids' Climate Suit*, LAW360, Jan. 30, 2017, https://www.law360.com/articles/886115.

[356] Motion for Leave to Appeal, *Juliana*, *supra* note 343 (D. Or. Mar. 7, 2017).

[357] Intervenor-Defendants' Motion for Certification of Order for Interloctory Appeal, *Juliana*, *supra* note 343 (D. Or. Mar. 10, 2017).

[358] *See, e.g.,* Respondents' Motion for Extension of Time to Respond to Motion to Sever and Consolidate, West Virginia, et al. v. EPA, No. 15-1363 (D.C. Cir. Mar. 1, 2017) (seeking additional time to respond to petitioners' motion to consolidate petitions for review of EPA's denial of petitions for reconsideration).

[359] For example, EPA's withdrawal and reconsideration of the Mid-Term Evaluation determination prompted Auto Alliance to dismiss their lawsuit challenging the determination. *See supra* "Light-Duty Motor Vehicles" section.

climate change lawsuits and legislation from Congress have the potential to reshape GHG regulation in the United States and, with it, American environmental law.

In: Climate Change
Editor: Iyana Hendricks
ISBN: 978-1-53617-757-2

Chapter 2

International Climate Change Assistance: Budget Authority, FY2009-FY2019*

Richard K. Lattanzio

The United States committed to providing financial assistance to developing countries for climate-change- related activities through the United Nations Framework Convention on Climate Change (UNFCCC). The United States ratified the UNFCCC in 1992 with the advice and consent of the Senate (U.S. Treaty Number: 102-38).

Among the obligations outlined in Article 4 of the UNFCCC, higher-income Parties (i.e., those listed in Annex II of the Convention, which were members of the Organization for Economic Cooperation and Development in 1992) sought to provide unspecified amounts of "financial resources, including for the transfer of technology, needed by developing countries to meet the agreed full incremental costs of implementing measures" to meet their general commitments under the UNFCCC. Further, "the

* This is an edited, reformatted and augmented version of Congressional Research Service Publication No. IF10397, updated November 25, 2019.

implementation of these commitments shall take into account the need for adequacy and predictability in the flow of funds and the importance of appropriate burden sharing among the developed country Parties."

Over the past several decades, and to varying degrees, the United States has delivered financial and technical assistance for climate-change-related activities in the developing world through a variety of bilateral and multilateral programs. (See *Table 1* for assistance provided over the past decade.)

U.S.-sponsored bilateral assistance has come through programs at the U.S. Agency for International Development (USAID); the Millennium Challenge Corporation; the Environmental Protection Agency; and the U.S. Departments of State, Agriculture, Commerce, and Energy, among others.

U.S.-sponsored multilateral assistance has come through contributions by the U.S. Departments of State and the Treasury to environmental funds at various international financial institutions and organizations such as the Global Environment Facility, the Green Climate Fund, the U.N. Development Program, the U.N. Environment Program, the UNFCCC's Special Climate Change Fund, the UNFCCC's Least Developed Country Fund, the World Bank's Climate Investment Funds, and the World Bank's Forest Carbon Partnership Facility, among others. Each fund has its own mission and particular capacities.

THE OBAMA ADMINISTRATION

Under President Barack Obama, the Administration aimed to integrate climate change considerations into relevant foreign assistance through a range of bilateral, multilateral, and private sector mechanisms to promote sustainable and resilient societies, foster low-carbon economic growth, and reduce greenhouse gas emissions from deforestation and land degradation.

To this end, President Obama signed the Presidential Policy Directive on Global Development (PPD-6), which called for the elevation of foreign development assistance as a national priority and outlined an integrated approach to development, diplomacy, and national security. One of the three main pillars of the directive was the Global Climate Change Initiative (GCCI). (The other two pillars were Feed the Future and the Global Health Initiative.) The GCCI was divided into three main programmatic initiatives or categories: (1) adaptation, (2) clean energy, and (3) sustainable landscapes.

The GCCI was funded primarily through programs at the State Department, the Department of the Treasury, and USAID. Funds for these programs were requested in the President's budget under the International Affairs Function 150 account for State, Foreign Operations, and Related Programs. Many GCCI activities were funded at agency subaccount levels, with allocations left to the discretion of the agencies under congressional consultation. Some additional international assistance was funded at other federal agencies. The Obama-era GCCI budget authority fluctuated between $900 million and $2 billion annually.

THE TRUMP ADMINISTRATION

Under President Donald Trump, the Administration has ceased providing financial assistance to developing countries for activities defined as related to climate change. The FY2018 Budget Blueprint, released on March 16, 2017, indicated that the administration would pursue a policy that "[e]liminates the Global Climate Change Initiative and fulfills the President's pledge to cease payments to the United Nations' (UN) climate change programs by eliminating U.S. funding related to the Green Climate Fund and its two precursor Climate Investment Funds." The Trump Administration has not requested funding for these programs in subsequent budget requests.

Table 1. International Climate Change Assistance, Budget Authority, FY2009-FY2019 (Nominal US$ in millions; n/a indicates "not available" or not reported)

Agency/Account	2009	2010	2011	2012	2013	2014	2015	2016	2017	2018	2019
Department of State	55	199	125	133	126	127	147	1095	n/a	n/a	n/a
Diplomatic and Consular Affairs	2	2	-	-	-	-	-	-	n/a	n/a	n/a
Economic Support Fund	24	158	89	96	91	92	112	1059*	n/a	n/a	n/a
International Organizations and Programs	29	39	36	37	35	36	36	36	n/a	n/a	n/a
USAID	222	383	398	349	335	351	331	305	n/a	n/a	n/a
Assistance for Europe, Eurasia, and Central Asia	15	26	-	15	-	-	-	16	n/a	n/a	n/a
Development Assistance	113	313	368	322	308	296	272	268	n/a	n/a	n/a
Economic Support Fund	94	44	30	12	27	55	59	21	n/a	n/a	n/a
Department of the Treasury	46	421	296	377	380	356	346	331	n/a	n/a	n/a
Debt Restructuring	20	20	16	12	11	-	-	-	-	-	-
Global Environment Facility	26	26	45	60	62	72	82	101	35	34	29
Clean Technology Fund	-	300	185	230	196	210	201	171	-	-	-
Strategic Climate Fund	-	75	50	75	110	75	63	60	-	-	-
Millennium Challenge Corporation	0	2	25	41	0	103	352	219	n/a	n/a	n/a
Other Agencies	50	75	85	59	76	77	76	78	n/a	n/a	n/a
Total	373	1080	929	959	917	1014	1252	2028	n/a	n/a	n/a

Source: CRS, from Office of Management and Budget, "Federal Expenditures on Science, Energy, and International Assistance Programs That Advance the Federal Response to Climate Change, Fiscal Years, 2013-2016," January 2017; U.S. Government Accountability Office, "Climate Change: Analysis of Reported Federal Funding," April 2018, GAO-18-223, and CRS correspondence with the Departments of State and the Treasury.

Notes: Numbers in the table may not sum due to rounding. "Other agencies" includes the Environmental Protection Agency; National Aeronautics and Space Administration; National Science Foundation; Peace Corps; U.S. Trade and Development Agency; and Departments of Agriculture, Commerce, and Energy.

* The Obama Administration made two contributions to the Green Climate Fund using FY2016 budget authority from the Economic Support Fund. Those contributions were for $500 million on March 8, 2016, and $500 million on January 17, 2017.

ISSUES FOR CONGRESS

Congress oversees U.S. government assistance to developing countries for climate-change-related initiatives. Congressional committees of jurisdiction have included the House Committees on Foreign Affairs, Financial Services, and Appropriations and the Senate Committees on Foreign Relations and Appropriations.

Congress undertakes several activities in regard to international climate change assistance, including (1) authorizing federal agency programs and multilateral fund contributions, (2) appropriating funds for those authorizations, (3) providing guidance to the agencies on authorized programs and appropriations, and (4) overseeing U.S. interests in the programs.

In: Climate Change
Editor: Iyana Hendricks
ISBN: 978-1-53617-757-2

Chapter 3

CLIMATE CHANGE: POTENTIAL ECONOMIC COSTS AND OPPORTUNITIES TO REDUCE FEDERAL FISCAL EXPOSURE*

Alfredo Gómez

WHY GAO DID THIS STUDY

Since 2005, federal funding for disaster assistance is at least $450 billion, including approximately $19.1 billion in supplemental appropriations signed into law on June 6, 2019. In 2018 alone, there were 14 separate billion-dollar weather and climate disaster events across the United States, with a total cost of at least $91 billion, according to the National Oceanic and Atmospheric Administration. The U.S. Global Change Research Program projects that disaster costs will likely increase

* This is an edited, reformatted and augmented version of the United States Government Accountability Office Testimony Before the Subcommittee on Environment, Committee on Oversight and Reform, House of Representatives, Publication No. GAO-20-338T, dated December 19, 2019.

as certain extreme weather events become more frequent and intense due to climate change.

The costs of recent weather disasters have illustrated the need for planning for climate change risks and investing in resilience. Resilience is the ability to prepare and plan for, absorb, recover from, and more successfully adapt to adverse events, according to the National Academies of Science, Engineering, and Medicine. Investing in resilience can reduce the need for far more costly steps in the decades to come.

Since February 2013, GAO has included *Limiting the Federal Government's Fiscal Exposure by Better Managing Climate Change Risks* on its list of federal program areas at high risk of vulnerabilities to fraud, waste, abuse, and mismanagement or most in need of transformation. GAO updates this list every 2 years. In March 2019, GAO reported that the federal government had not made measurable progress since 2017 to reduce fiscal exposure to climate change.

This testimony—based on reports GAO issued from October 2009 to October 2019—discusses 1) what is known about the potential economic effects of climate change in the United States and the extent to which this information could help federal decision makers manage climate risks across the federal government, (2) the fiscal exposure facing the federal government due to climate risks and current efforts to address that exposure, (3) the extent to which the federal government has invested in resilience to climate change impacts, and (4) how the federal government could reduce fiscal exposure to the effects of climate change.

GAO had made 62 recommendations related to the Limiting the Federal Government's Fiscal Exposure by Better Managing Climate Change Risks high-risk area. As of December 2018, 25 of those recommendations remained open.

WHAT GAO FOUND

The estimated economic effects of climate change, while imprecise, can convey useful insight about potential damages in the United States. In

September 2017, GAO reported that the potential economic effects of climate change could be significant and unevenly distributed across sectors and regions (see figure). This is consistent with the 2018 findings of the U.S. Global Change Research Program's Fourth National Climate Assessment, which concluded, among other things, that the continued increase in the frequency and extent of high-tide flooding due to sea level rise threatens America's trillion-dollar coastal infrastructure.

Information about the potential economic effects of climate change could inform decision makers about significant potential damages in different U.S. sectors or regions. According to prior GAO work, this information could help decision makers identify significant climate risks as an initial step toward managing them.

The federal government faces fiscal exposure from climate change risks in several areas, including:

- *Disaster aid:* due to the rising number of natural disasters and increasing reliance on federal assistance. GAO has previously reported that the federal government's fragmented and reactive approach to funding disaster resilience presented challenges to effective reduction of climate-related risks. GAO has also reported that, due to an artificially low indicator for determining a jurisdiction's ability to respond to disasters that was set in 1986, the Federal Emergency Management Agency risks recommending federal assistance for jurisdictions that could recover on their own.
- *Federal insurance for property and crops:* due, in part, to the vulnerability of insured property and crops to climate change impacts. Federal flood and crop insurance programs were not designed to generate sufficient funds to fully cover all losses and expenses. The flood insurance program, for example, was about $21 billion in debt to the Treasury as of April 2019. Further, the Congressional Budget Office estimated in May 2019 that federal crop insurance would cost the federal government an average of about $8 billion annually from 2019 through 2029.

- *Operation and management of federal property and lands:* due to the hundreds of thousands of federal facilities and millions of acres of land that could be affected by a changing climate and more frequent extreme events. For example, in 2018, Hurricane Michael devastated Tyndall Air Force Base in Florida, with a preliminary repair estimate of $3 billion.

As we reported in October 2019, our past work shows an absence of government-wide strategic planning for climate change. Specifically, our past work has identified limitations related to strategic planning for climate change that includes a lack of coordination, prioritization, and consolidation of strategic priorities. In our March 2019 High-Risk Update, we assessed the federal government's progress since 2017 related to climate change strategic planning against five criteria and found that the federal government had not met any of the criteria for removal from the high-risk list.

Federal investments in resilience to reduce fiscal exposures have been limited. As GAO has reported, enhancing resilience can reduce fiscal exposure by reducing or eliminating long-term risk to people and property from natural hazards. For example, a 2018 interim report by the National Institute of Building Sciences estimated approximate benefits to society in excess of costs for several types of resilience projects. While precise benefits are uncertain, the report estimated that for every dollar invested in designing new buildings to particular design standards, society could accrue benefits amounting to about $11 on average.

GAO's March 2019 High-Risk report identified a number of recommendations GAO has made related to fiscal exposure to climate change. The federal government could reduce its fiscal exposure by implementing these recommendations. Among GAO's key government-wide recommendations are:

- Entities within the Executive Office of the President (EOP) should work with partners to establish federal strategic climate change priorities that reflect the full range of climate-related federal activities;

- Entities within EOP should use information on potential economic effects from climate change to help identify significant climate risks and craft appropriate federal responses;
- Entities within EOP should designate a federal entity to develop and update a set of authoritative climate observations and projections for use in federal decision making, and create a national climate information system with defined roles for federal agencies and certain nonfederal entities; and
- The Department of Commerce should convene federal agencies to provide the best-available forward-looking climate information to organizations that develop standards and building codes to enhance infrastructure resilience.

Further, in October 2019, GAO reported that Congress could consider establishing a federal organizational arrangement to periodically identify and prioritize climate resilience projects for federal investment. GAO also issued the Disaster Resilience Framework to serve as a guide for analysis of federal action to facilitate and promote resilience to natural disasters, including resilience to climate change.

Chairman Rouda, Ranking Member Comer, and Members of the Subcommittee:

Thank you for the opportunity to discuss our work on how to limit the federal government's fiscal exposure by better managing climate change risks, an area that has been on our High-Risk List since February 2013.[1] Addressing climate change risks requires advanced planning and investment to reduce the need for far more costly steps in the decades to come, which, as we have previously reported, the federal government is not well organized to do. The costs associated with recent disasters have

[1] Our High-Risk List identifies federal program areas that are at high risk of vulnerabilities to fraud, waste, abuse, and mismanagement or most in need of transformation. See GAO, *High-Risk Series: An Update*, GAO-13-283 (Washington, D.C.: February 2013).

illustrated the need for such planning and investment. In 2018 alone, there were 14 separate billion-dollar weather and climate disaster events across the United States, with a total cost of at least $91 billion, according to the National Oceanic and Atmospheric Administration (NOAA).[2] Further, on June 6, 2019, a supplemental appropriation of approximately $19.1 billion was signed into law for recent disasters.[3]

The U.S. Global Change Research Program (USGCRP), which coordinates and integrates the activities of 13 federal agencies that research changes in the global environment and their implications for society, reported in its November 2018 Fourth National Climate Assessment that climate change is playing a role in the increasing frequency of some types of extreme weather that lead to the billion-dollar disasters.[4] These changes include the rise in vulnerability to drought, lengthening wildfire seasons, and potential for extremely heavy rainfall becoming more common in some regions. USGCRP reported in the prior assessment that the costs of many of these disasters will likely increase as extreme weather events become more frequent and intense with climate change.[5]

In my testimony today, I will discuss (1) what is known about the potential economic effects of climate change in the United States and the extent to which this information could help federal decision makers manage climate risks across the federal government, (2) the fiscal exposure facing the federal government due to climate risks and current efforts to address that exposure, (3) the extent to which the federal government has

---

[2] National Oceanic and Atmospheric Administration, National Centers for Environmental Information, *U.S. Billion-Dollar Weather and Climate Disasters (2019)*, accessed June 3, 2019, https://www.ncdc.noaa.gov/billions/time-series.

[3] Pub. L. No. 116-20, 133 Stat. 871 (2019).

[4] D.R. Reidmiller, C.W. Avery, D. R. Easterling, K. E. Kunkel, K. L. M. Lewis, T. K. Maycock, and B. C. Stewart (eds.), *2018: Impacts, Risks, and Adaptation in the United States: Fourth National Climate Assessment, Volume II* (Washington, D.C.: U.S. Global Change Research Program, November 2018). Under the Global Change Research Act of 1990 (Pub. L. No. 101-606, § 103 (1990)), USGCRP is to periodically prepare a scientific assessment—known as the National Climate Assessment—which is an important resource for understanding and communicating climate change science and impacts in the United States. The Office of Science and Technology Policy within the Executive Office of the President oversees USGCRP.

[5] Jerry M. Melillo, Terese (T.C.) Richmond, and Gary W. Yohe, eds., *Climate Change Impacts in the United States: The Third National Climate Assessment* (Washington, D.C.: U.S. Global Change Research Program, May 2014).

invested in resilience to climate change impacts,[6] and (4) how the federal government could reduce fiscal exposure to the effects of climate change. My testimony is based on reports we issued from October 2009 to October 2019. More detailed information on our objectives, scope, and methodology can be found in those reports.

The work upon which this statement is based was conducted in accordance with generally accepted government auditing standards. Those standards require that we plan and perform the audit to obtain sufficient, appropriate evidence to provide a reasonable basis for our findings and conclusions based on our audit objectives. We believe that the evidence obtained provides a reasonable basis for our findings and conclusions based on our audit objectives.

Information on the Potential Economic Effects of Climate Change in the United States Could Help Federal Decision Makers Better Manage Climate Risks

We reported in September 2017 that while estimates of the economic effects of climate change are imprecise due to modeling and information limitations, they can convey useful insight into broad themes about

---

[6] The National Academies of Sciences, Engineering, and Medicine define resilience as the ability to prepare and plan for, absorb, recover from, and more successfully adapt to adverse events. See the National Academies, Committee on Increasing National Resilience to Hazards and Disasters and Committee on Science, Engineering, and Public Policy, *Disaster Resilience: A National Imperative* (Washington, D.C.: 2012). We reported in May 2016 that two related sets of actions can enhance resilience by reducing risk: climate change adaptation and predisaster hazard mitigation. Adaptation is defined as adjustments to natural or human systems in response to actual or expected climate change. Predisaster hazard mitigation refers to actions taken to reduce the loss of life and property by lessening the impacts of adverse events and applies to all hazards, including terrorism and natural hazards, such as health pandemics or weather-related disasters. In this testimony, we use "resilience" for consistency and to encompass both of these sets of actions as they relate to addressing climate risks. GAO, *Climate Change: Selected Governments Have Approached Adaptation through Laws and Long-Term Plans*, GAO- 16-454 (Washington, D.C.: May 12, 2016).

potential damages in the United States.[7] We also reported that according to the two national-scale studies available at the time that examined the economic effects of climate change across U.S. sectors, potential economic effects could be significant and these effects will likely increase over time for most of the sectors analyzed.[8] For example, for 2020 through 2039, one of the studies estimated from $4 billion to $6 billion in annual coastal property damages from sea level rise and more frequent and intense storms.[9]

In addition, the national-scale studies we reviewed and several experts we interviewed for the September 2017 report suggested that potential economic effects could be unevenly distributed across sectors and regions. For example, one of the studies estimated that the Southeast, Midwest, and Great Plains regions will likely experience greater combined economic effects than other regions, largely because of coastal property damage in the Southeast and changes in crop yields in the Midwest and Great Plains (see figure 1).[10] This is consistent with the findings of the Fourth National

---

[7] GAO, Climate Change: Information on Potential Economic Effects Could Help Guide Federal Efforts to Reduce Fiscal Exposure, GAO-17-720 (Washington, D.C.: Sept. 28, 2017).

[8] These national-scale studies were the Environmental Protection Agency's (EPA) *Climate Change Impacts and Risk Analysis*—a summary study of an ongoing EPA project—and the Rhodium Group's *American Climate Prospectus.* See Environmental Protection Agency, Office of Atmospheric Programs, *Climate Change in the United States: Benefits of Global Action*, EPA 430-R-15-001 (Washington, D.C.: 2015). The EPA project on which the summary study was based was coordinated by EPA's Office of Atmospheric Programs—Climate Change Division, with contributions from national laboratories and the academic and private sectors. The detailed methods and results of the project were published in a 2014 special issue of the peer-reviewed journal, *Climatic Change* entitled, "A Multi-Model Framework to Achieve Consistent Evaluation of Climate Change Impacts in the United States." An update to this project was used in the 2018 Fourth National Climate Assessment. Also see Rhodium Group, LLC, *American Climate Prospectus: Economic Risks in the United States* (New York: October 2014). The *American Climate Prospectus* was funded by the Risky Business Project, a project funded by Bloomberg Philanthropies, the Paulsen Institute, and TomKat Charitable Trust; the Skoll Global Threats Fund; and the Rockefeller Family Fund. The Rhodium Group, LLC, a research consultancy and advisory company, coordinated the effort, which involved authors from universities and the private sector. This study was later published by the Columbia University Press in 2015: Trevor Houser et al., *Economic Risks of Climate Change: An American Prospectus* (New York: Columbia University Press, 2015). An update to this analysis was published in *Science* in June 2017: Solomon Hsiang et al. "Estimating Economic Damage from Climate Change in the United States," *Science*, vol. 356(2017).

[9] Rhodium Group, *American Climate Prospectus.*

[10] Rhodium Group, *American Climate Prospectus.*

Climate Assessment.[11] For example, according to that assessment, the continued increase in the frequency and extent of high- tide flooding due to sea level rise threatens America's trillion-dollar coastal property market and public infrastructure sector.

As we reported in September 2017, information on the potential economic effects of climate change could help federal decision makers better manage climate risks, according to leading practices for climate risk management, economic analysis we reviewed, and the views of several experts we interviewed.[12] For example, such information could inform decision makers about significant potential damages in different U.S. sectors or regions. According to several experts and our prior work, this information could help federal decision makers identify significant climate priorities as an initial step toward managing climate risks.[13] Such a first step is consistent with leading practices for climate risk management and federal standards for internal control.[14] For example, leading practices from the National Academies call for climate change risk management efforts that focus on where immediate attention is needed.[15] As noted in our September 2017 report, according to a 2010 National Academies report, other literature we reviewed, and several experts we interviewed, to make informed choices, decision makers need more comprehensive information on economic effects to better understand the potential costs of climate change to society and begin to develop an understanding of the benefits and costs of different options for managing climate risks.[16]

---

[11] D.R. Reidmiller et al., *Fourth National Climate Assessment, Volume II.*

[12] In that report, we also found that additional economic information could help federal, state, local, and private sector decision makers manage climate risks that drive federal fiscal exposure. GAO-17-720.

[13] GAO-17-720.

[14] National Research Council of the National Academies, America's Climate Choices: Panel on Adapting to the Impacts of Climate Change, Adapting to the Impacts of Climate Change (Washington, D.C.: 2010), and GAO, Standards for Internal Control in the Federal Government, GAO-14-704G (Washington, D.C.: September 2014).

[15] National Research Council of the National Academies, America's Climate Choices: Panel on Adapting to the Impacts of Climate Change, Adapting to the Impacts of Climate Change.

[16] GAO-17-720.

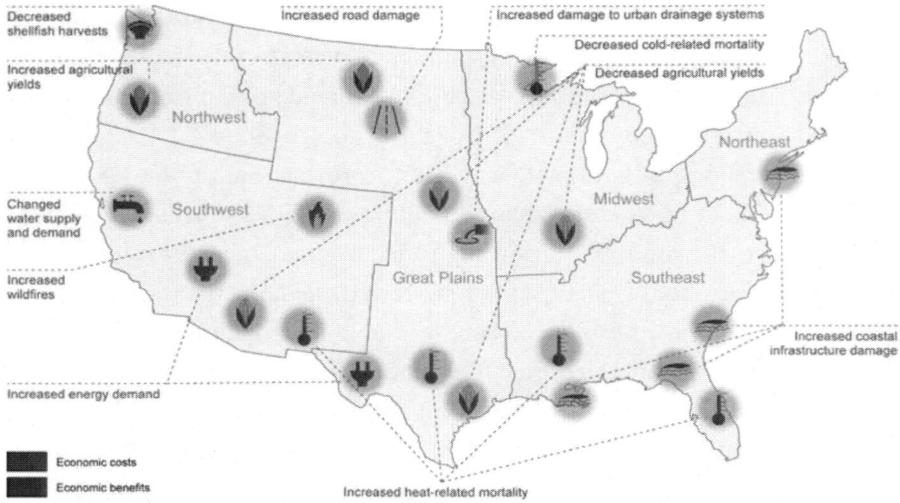

Source: GAO analysis of Environmental Protection Agency. *Climate Change Impacts in the United States: Benefits of Global Action* (Washington, D.C.: 2015), and Solomon Hsiang et al., "Estimating Economic Damage from Climate Change in the United States." Science, vol. 356 (2017): Map Resources (map). | GAO-20-338T.

Figure 1. Examples of Potential Economic Effects from Climate Change by 2100.

THE FEDERAL GOVERNMENT FACES FISCAL EXPOSURE FROM CLIMATE CHANGE RISKS, BUT OUR PAST WORK SHOWS AN ABSENCE OF GOVERNMENT-WIDE STRATEGIC PLANNING

The federal government faces fiscal exposure from climate change risks in a number of areas, and this exposure will likely increase over time, as we concluded in September 2017.[17] In the March 2019 update to our High-Risk List, we summarized our previous work that identified several

[17] GAO-17-720.

of these areas across the federal government, including programs related to the following:[18]

- *Disaster aid.* The rising number of natural disasters and increasing reliance on federal assistance are a key source of federal fiscal exposure, and this exposure will likely continue to rise. Since 2005, federal funding for disaster assistance has been at least $450 billion.[19] In September 2018, we reported that four hurricane and wildfire disasters in 2017 created an unprecedented demand for federal disaster resources and that Hurricanes Harvey, Irma, and Maria ranked among the top five costliest hurricanes on record.[20] Subsequently, the fall of 2018 brought additional catastrophic disasters such as Hurricanes Florence and Michael and devastating California wildfires, with further needs for federal disaster assistance. Disaster costs are projected to increase as certain

---

[18] We have identified other areas with potential links to climate and the federal budget in past reports, including global migration, state and local infrastructure, federal supply chains, and public health. See GAO, *Climate Change: Activities of Selected Agencies to Address Potential Impact on Global Migration*, GAO-19-166 (Washington, D.C.: Jan. 17, 2019); *Climate Information: A National System Could Help Federal, State, Local, and Private Sector Decision Makers Use Climate Information*, GAO-16-37 (Washington, D.C.: Nov. 23, 2015); *Federal Supply Chains: Opportunities to Improve the Management of Climate-Related Risks*, GAO-16-32 (Washington, D.C.: Oct. 13, 2015); and *Climate Change: HHS Could Take Further Steps to Enhance Understanding of Public Health Risks*, GAO-16-122 (Washington, D.C.: Oct. 5, 2015). We also have ongoing work in many areas related to federal fiscal exposure to climate change, examining issues such as how to identify and prioritize resilience projects to build resilience to climate change impacts, how to make water infrastructure more resilient to the impacts of climate change, and how to help communities voluntarily relocate to avoid climate change impacts.

[19] This total includes, for fiscal years 2005 through 2014, $278 billion that GAO found that the federal government had obligated for disaster assistance. See GAO, *Federal Disaster Assistance: Federal Departments and Agencies Obligated at Least $277.6 Billion during Fiscal Years 2005 through 2014*, GAO-16-797 (Washington, D.C.: Sept. 22, 2016). It also includes, for fiscal years 2015 through 2018, $124 billion in select supplemental appropriations to federal agencies for disaster assistance, with approximately $7 billion in annual appropriations to the Disaster Relief Fund (a total of $28 billion for the 4-year period). For fiscal years 2015 through 2018, it does not include other annual appropriations to federal agencies for disaster assistance. Lastly, on June 6, 2019, the Additional Supplemental Appropriations for Disaster Relief Act of 2019 was signed into law, which provides approximately $19.1 billion for disaster assistance. Pub. L. No. 116- 20, 133 Stat. 871 (2019).

[20] GAO, *2017 Hurricanes and Wildfires: Initial Observations on the Federal Response and Key Recovery Challenges*, GAO-18-472 (Washington, D.C.: Sept. 4, 2018).

extreme weather events become more frequent and intense due to climate change—as USGCRP observed and projected.[21] We reported in July 2015 that the federal government's fragmented and reactive approach to funding disaster resilience presented challenges to effective reduction of climate-related risks.[22] In addition, our prior work found that the Federal Emergency Management Agency's (FEMA) primary indicator for determining whether to recommend that a jurisdiction receive disaster assistance—which was set in 1986—is artificially low because it does not accurately reflect the ability of state and local governments to respond to disasters.[23] Without an accurate assessment of a jurisdiction's capability to respond to a disaster without federal assistance, we found that FEMA runs the risk of recommending that the President award federal assistance to jurisdictions that have the capability to respond and recover on their own.

- *Federal insurance for property and crops.* The National Flood Insurance Program (NFIP) and the Federal Crop Insurance Corporation are sources of federal fiscal exposure due, in part, to

[21] Melillo et al., Climate Change Impacts in the United States: The Third National Climate Assessment.

[22] For example, from fiscal years 2011 to 2014, the Federal Emergency Management Agency obligated more than $3.2 billion for the Hazard Mitigation Grant Program for postdisaster hazard mitigation while obligating approximately $222 million for the Pre- Disaster Mitigation Grant Program. GAO, *Hurricane Sandy: An Investment Strategy Could Help the Federal Government Enhance National Resilience for Future Disasters*, GAO-15- 515 (Washington, D.C.: July 30, 2015).

[23] GAO, *Federal Disaster Assistance: Improved Criteria Needed to Assess a Jurisdiction's Capability to Respond and Recover on Its Own*, GAO-12-838 (Washington, D.C.: Sept. 12, 2012). We recommended, among other things, that FEMA develop and implement a methodology that more comprehensively assesses a jurisdiction's capacity to respond to and recover from a disaster without federal assistance, including fiscal capacity and consideration of response and recovery capabilities. The Disaster Recovery Reform Act of 2018 includes a provision directing the FEMA Administrator to initiate rulemaking to update the factors considered when evaluating requests for major disaster declarations. Pub. L. No. 115-254, div. D, § 1239, 132 Stat. at 3466 (2018). According to FEMA documentation, as of September 2019, the agency was working to implement this provision through rulemaking proposals, including increasing the per capita indicator to account for inflation.

the vulnerability of insured property and crops to climate change.[24] These programs provide coverage where private markets for insurance do not exist, typically because the risk associated with the property or crops is too great to privately insure at a cost that buyers are willing to accept. From 2013 to 2017, losses paid under NFIP and the federal crop insurance program totaled $51.3 billion.[25] Federal flood and crop insurance programs were not designed to generate sufficient funds to fully cover all losses and expenses, which means the programs need budget authority from Congress to operate. NFIP, for example, was about $21 billion in debt to the Department of the Treasury as of April 2019.[26] Further, the Congressional Budget Office estimated in May 2019 that federal crop insurance would cost the federal government an average of about $8 billion annually from 2019 through 2029.[27]

- *Operation and management of federal property and lands.* The federal government owns and operates hundreds of thousands of facilities and manages millions of acres of land that could be affected by a changing climate and represent a significant federal fiscal exposure. For example, the Department of Defense (DOD) owns and operates domestic and overseas infrastructure with an estimated replacement value of about $1 trillion. In September 2018, Hurricane Florence damaged Camp Lejeune and other Marine Corps facilities in North Carolina, resulting in a preliminary Marine Corps repair estimate of $3.6 billion. One month later, Hurricane Michael devastated Tyndall Air Force Base in Florida, resulting in a preliminary Air Force repair estimate of

---

[24] NFIP is administered by FEMA within the U.S. Department of Homeland Security, and the Federal Crop Insurance Corporation is administered by the Risk Management Agency within the U.S. Department of Agriculture.

25 This figure is based on FEMA and Risk Management Agency published data. This does not include the costs of running these programs or the premiums collected to partially offset the costs. Losses for the crop insurance program are losses associated with crops harvested in that year, also known as crop year.

[26] Department of the Treasury, Bureau of the Fiscal Service, *Monthly Treasury Statement. Table 6, Schedule C* (Washington, D.C.: April 2019).

[27] Congressional Budget Office, CBO's May 2019 Baseline for Farm Programs (Washington, D.C.: May 2, 2019).

> $3 billion and upwards of 5 years to complete the work. In addition, we recently reported that the federal government manages about 650 million acres of land in the United States that could be vulnerable to climate change, including the possibility of more frequent and severe droughts and wildfires.[28] Appropriations for federal wildland fire management activities have increased considerably since the 1990s, as we and the Congressional Research Service have reported.[29]

As we reported in October 2019, our past work shows an absence of government-wide strategic planning for climate change.[30] Specifically, our past work identifies limitations related to strategic planning for climate change that include a lack of coordination, prioritization, and consolidation of strategic priorities. For example, we reported in October 2009 that the federal government's emerging climate resilience activities were carried out in an ad hoc manner and were not well coordinated across federal agencies.[31] In May 2011, we reported that federal officials did not have a shared understanding of strategic government-wide priorities related to climate change.[32] In the same report, we found that there was not a consolidated set of strategic priorities integrating climate change programs and activities across the federal government.

---

[28] GAO, *Climate Change: Various Adaptation Efforts Are Under Way at Key Natural Resource Management Agencies*, GAO-13-253 (Washington, D.C.: May 31, 2013).

[29] GAO, Budget Issues: Opportunities to Reduce Federal Fiscal Exposures Through Greater Resilience to Climate Change and Extreme Weather, GAO-14-504T (Washington, D.C.: July 29, 2014), and Congressional Research Service, Wildfire Suppression Spending: Background, Issues, and Legislation in the 115th Congress, R44966 (Washington, D.C.: Nov. 8, 2017).

[30] GAO, *Climate Resilience: A Strategic Investment Approach for High-Priority Projects Could Help Target Federal Resources,* GAO-20-127 (Washington, D.C.: Oct. 23, 2019). Several federal agencies prepared climate change adaptation plans that outline strategies to reduce the vulnerability of federal programs, assets, and investments to the impacts of climate change. These plans were prepared in response to the 2013 Executive Order 13653, *Preparing the United States for the Impacts of Climate Change*, which has been rescinded. These plans were agency specific and do not represent a government-wide strategic planning effort for climate change.

[31] GAO, Climate Change Adaptation: Strategic Federal Planning Could Help Government Officials Make More Informed Decisions, GAO-10-113 (Washington, D.C.: Oct. 7, 2009).

[32] GAO, Climate Change: Improvements Needed to Clarify National Priorities and Better Align Them with Federal Funding Decisions, GAO-11-317 (Washington, D.C.: May 20, 2011).

GAO's High-Risk List

Since the early 1990s, GAO's high-risk list has called attention to agencies and program areas that are at high risk due to their vulnerabilities to fraud, waste, abuse, and mismanagement, or are most in need of transformation. Every 2 years at the start of a new Congress, GAO reevaluates agency progress in addressing issues on the high-risk list against five criteria to determine if progress has been made. The criteria are (1) leadership commitment to address the risk, (2) agency capacity to resolve the risk, (3) a corrective action plan for addressing the risk, (4) a program to monitor the effectiveness of corrective measures, and (5) ability to demonstrate progress in resolving the high-risk area.

- **Leadership commitment:** not met. A March 2017 Executive Order (E.O. 13783) revoked policies, including the Climate Action Plan and a 2013 Executive Order (E.O. 13653), that we previously found had demonstrated leadership support for reducing aspects of fiscal exposure to climate change.[a] E.O. 13653, which had directed agencies to submit resilience plans to Council on Environmental Quality and the Office of Management and Budget (OMB) for review, and E.O. 13834 revoked E.O. 13693, which had directed OMB to evaluate agencies' strategic sustainability performance.
- **Capacity:** not met. A May 2018 Executive Order (E.O.13834) revoked a 2013 Executive Order (E.O. 13693), which we had previously found partially met this criterion.[b] Specifically, E.O. 13693 had directed the Office of Personnel Management to include resilience in federal training and also directed the creation of interagency workgroups to address, among other things, resilience planning in coordination with states and other stakeholders.
- **Action plan:** not met. The two aforementioned executive orders (13783 and 13834) withdrew previous direction to develop resilience plans and strategic sustainability performance plans that identified agency actions to prepare for climate change impacts and improve resilience.[c]
- **Monitoring:** not met. Executive orders 13783 and 13834 also eliminated monitoring mechanisms that we previously found had partially met this criterion. Specifically, E.O. 13783 revoked

- **Demonstrating progress:** not met. The federal government has not implemented key recommendations in this area. This includes (1) a May 2011 recommendation to develop a strategic plan to guide the nation's efforts to adapt to climate change that includes clear federal priorities that reflect the full range of climate-related federal activities and that establishes clear roles, responsibilities, and working relationships among federal, state, and local governments; (2) a September 2017 recommendation to use information on potential economic effects from climate change to help identify significant climate risks and craft appropriate federal responses; and (3) an April 2018 recommendation to provide information on fiscal exposures related to climate change to Congress in conjunction with future reports on climate change funding.[d] As of our 2019 update to our high-risk list, the Executive Office of the President and OMB had not implemented any of these recommendations.[e]

Source: GAO-19-157SP (GAO High Risk Series: Substantial Efforts Needed to Achieve Greater Progress on High-Risk Areas). | GAO-20-338T.

[a] Executive Order 13783, Promoting Energy Independence and Economic Growth (Mar. 28, 2017). Executive Order 13653, Preparing the United States for the Impacts of Climate Change (revoked) (Nov. 6, 2013).

[b] Executive Order 13834, Efficient Federal Operations (May 17, 2018). Executive Order 13693, Planning for Federal Sustainability in the Next Decade (revoked) (Mar. 19, 2015).

[c] The Mitigation Framework Leadership Group, an intergovernmental coordinating body, finalized the National Mitigation Investment Strategy in August 2019. However, as noted, our review of the strategy indicates that it does not include a detailed strategic approach to prioritize investments for disaster risk reduction that explicitly accounts for future climate change risks. According to Federal Emergency Management Agency officials, the strategy sets goals and recommendations that set the stage for developing approaches to address changing conditions.

[d] GAO, Climate Change: Improvements Needed to Clarify National Priorities and Better Align Them with Federal Funding Decisions, GAO-11-317 (Washington, D.C.: May 20, 2011); Climate Change: Information on Potential Economic Effects Could Help Guide Federal Efforts to Reduce Fiscal Exposure, GAO-17-720 (Washington, D.C.: Sept. 28, 2017); and Climate Change: Analysis of Reported Federal Funding, GAO-18-223 (Washington, D.C.: Apr. 30, 2018).

[e] GAO, High-Risk Series: Substantial Efforts Needed to Achieve Greater Progress on High- Risk Areas, GAO-19-157SP (Washington, D.C.: Mar. 6, 2019).

Figure 2. Federal Progress since 2017 Related to Climate Change Strategic Planning.

In our March 2019 High-Risk Update, we reported that one area of government-wide action needed to reduce federal fiscal exposure is in the federal government's role as the leader of a strategic plan that coordinates federal efforts and informs state, local, and private sector action.[33] For our 2019 High-Risk Update, we assessed the federal government's progress since 2017 related to climate change strategic planning against five criteria and found that the federal government had not met any of the criteria for removal from the high-risk list.[34] Specifically, since our 2017 high-risk update, four ratings regressed to "not met" and one remained unchanged as "not met." (See fig. 2.) We have made 62 recommendations related to the climate change high-risk area, 17 of which address improving federal climate change strategic planning. As of August 2019, no action had been taken toward 14 of those 17 recommendations—one dating back to 2003.[35]

Federal Investments in Resilience to Climate Change Impacts Have Been Limited

Although the federal government faces fiscal exposure to climate change, its investments in resilience to climate change impacts have been limited. One way to reduce federal fiscal exposure is to enhance resilience by reducing or eliminating long-term risk to people and property from natural hazards. For example, in September 2018 we reported that elevated homes and strengthened building codes in Texas and Florida prevented

---

[33] GAO, High-Risk Series: Substantial Efforts Needed to Achieve Greater Progress on High-Risk Areas, GAO-19-157SP (Washington, D.C.: Mar. 6, 2019).

[34] Every 2 years, at the start of a new Congress, GAO reevaluates agency progress in addressing issues on the High-Risk List against five criteria to determine if progress has been made. The criteria are (1) leadership commitment to address the risk, (2) agency capacity to resolve the risk, (3) a corrective action plan to addressing the risk, (4) a program to monitor the effectiveness of corrective measures, and (5) ability to demonstrate progress in resolving the high-risk area. GAO-13-283.

[35] Based on the most current information available as of August 2019, of the 17 recommendations related to strategic planning, 11 were closed as not implemented, three were closed as implemented, and three remain open.

greater damages during the 2017 hurricane season.[36] In addition, one company participating in a 2014 forum we held on preparing for climate-related risks noted that for every dollar it invested in resilience efforts, the company could prevent $5 in potential losses.[37] Finally, a 2018 interim report by the National Institute of Building Sciences examined a sample of federal grants for hazard mitigation. The interim report estimated approximate benefits to society (i.e., homeowners and communities) in excess of costs for several types of resilience projects through the protection of lives and property, and prevention of other losses, though precise benefits are uncertain.[38] According to the interim report, for every grant dollar the federal government spent on resilience projects, over time, society is estimated to accrue benefits amounting to the following:

- About $3 on average from projects addressing the effects of fire in the wildland urban interface, with most benefits (approximately 70 percent) coming from the protection of property (i.e., avoiding property losses).
- About $5 on average from projects to address hurricane-force and tornado-force winds, with most benefits (approximately 90 percent) coming from the protection of lives. This includes

---

[36] Specifically, FEMA officials said Hurricane Harvey demonstrated how prior hazard mitigation projects prevented greater damages (e.g., elevated homes and equipment sustained less damage). FEMA officials said Florida strengthened its building codes for resilience as a result of Hurricane Andrew in 1992 and Hurricane Matthew in 2016. GAO- 18-472.

[37] GAO, *Highlights of a Forum: Preparing for Climate-Related Risks: Lessons from the Private Sector*, GAO-16-126SP (Washington, D.C.: Nov. 19, 2015).

[38] Benefit estimates from federal grants convey the magnitude of potential long-term benefits to society, primarily homeowners and local residents, and are not precise estimates. This interim report analyzes a small sample of projects for high-risk buildings selected with specific criteria to examine hazard mitigation grants awarded by FEMA, the Economic Development Administration, and the Department of Housing and Urban Development from 1993 to 2016 to address various hazards. Extrapolation to a broader set of grants needs to be interpreted in the context of the selected sample. These hazards included fires at the wildland-urban interface (i.e., fires in areas where homes are built near or among lands prone to wildland fire), hurricane-force and tornado-force winds, and riverine floods (i.e., floods that occur when river flows exceed the capacity of the river channel). See Multihazard Mitigation Council, a council of the National Institute of Building Sciences, *Natural Hazard Mitigation Saves: 2018 Interim Report* (Washington, D.C.: December 2018).

avoiding deaths, nonfatal injuries, and causes of posttraumatic stress.

- About $7 on average from projects that buy out buildings prone to riverine flooding, with most benefits (approximately 65 percent) coming from the protection of property.
- The interim report also projected that society could accrue benefits amounting to about $11 on average for every dollar invested in designing new buildings to meet the 2018 International Building Code and the 2018 International Residential Code—the model building codes that the International Code Council developed—with most benefits (46 percent) coming from the protection of property.[39]

We reported in October 2009 that the federal government's activities to build resilience to climate change were carried out in an ad hoc manner and were not well coordinated across federal agencies.[40] We reported similar findings in October 2019.[41] Federal agencies have included some of these activities within existing programs and operations—a concept known as mainstreaming. For example, the Fourth National Climate Assessment reported that the U.S. military integrates climate risks into its analysis, plans, and programs, with particular attention paid to climate effects on force readiness, military bases, and training ranges.[42] However, according

[39] The benefit-cost ratios cited in the interim report are based on a relatively narrow set of disaster-loss data, and the report is not comprehensive. Benefits are estimated using many assumptions at all stages of the analysis and are subject to a high degree of model uncertainty and sensitivity. The International Code Council is a member-focused association with over 64,000 members dedicated to developing model codes and standards for use in the design, build, and compliance process to construct safe, sustainable, affordable, and resilient structures. The report used a baseline of buildings constructed to a prior generation of codes represented by 1990s-era design and NFIP requirements to examine code-related mitigation strategies for floods, winds, and earthquakes.

[40] GAO-10-113.

[41] GAO-20-127.

[42] R. Lempert, J. Arnold, R. Pulwarty, K. Gordon, K. Greig, C. Hawkins Hoffman, D. Sands, and C. Werrell, "2018: Reducing Risks Through Adaptation Actions," in *Impacts, Risks, and Adaptation in the United States: Fourth National Climate Assessment, Volume II* (Washington, D.C.: U.S. Global Change Research Program, 2018). We also reported in May 2014 that officials from the Office of the Secretary of Defense and the military departments stated that their goal is to address potential climate change impacts and

to the Fourth National Climate Assessment, while a significant portion of climate risk can be addressed by mainstreaming, the practice may reduce the visibility of climate resilience relative to dedicated, stand-alone approaches and may prove insufficient to address the full range of climate risks.[43]

In addition, as we reported in March 2019, the Disaster Recovery Reform Act of 2018 (DRRA) was enacted in October 2018 and could improve state and local resilience to disasters. DRRA, among other things, allows the President to set aside, with respect to each major disaster, a percentage of the estimated aggregate amount of certain grants to use for predisaster hazard mitigation and makes federal assistance available to state and local governments for building code administration and enforcement.[44] However, it is too early to tell what impact implementing the act will have on state and local resilience.

The federal government has made some limited investments in resilience, and DRRA could enable additional improvements at the state and local levels. However, we reported in October 2019 that the federal government does not have a strategic approach for investing in climate resilience projects—that is, an intentional, crosscutting approach in which the federal government identifies and prioritizes projects for the purpose of enhancing climate resilience.[45] Federal agencies may take actions to invest in projects with potential climate resilience benefits related to their own mission areas using funds from federal programs designed for other

---

vulnerabilities through existing infrastructure planning processes so that the effects of climate change are considered in the same way other impacts and vulnerabilities—such as force protection—are currently considered. GAO, *Climate Change Adaptation: DOD Can Improve Infrastructure Planning and Processes to Better Account for Potential Impacts*, GAO-14-446 (Washington, D.C.: May 30, 2014).

[43] Lempert, R., J. Arnold, R. Pulwarty, K. Gordon, K. Greig, C. Hawkins Hoffman, D. Sands, and C. Werrell, 2018: Reducing Risks Through Adaptation Actions. *In Impacts, Risks, and Adaptation in the United States: Fourth National Climate Assessment, Volume II* (Washington, D.C.: U.S. Global Change Research Program, 2018).

[44] FAA Reauthorization Act of 2018, Pub. L. No. 115-254, div. D, §§ 1206(a)(3), 1234(a)(2)(C), 1234(a)(5), 132 Stat. 3186, 3440, 3462 (2018). The FAA Reauthorization Act of 2018, which included DRRA, became law on October 5, 2018.

[45] GAO-20-127. To make this assessment, we reviewed documents describing federal climate change activities, past GAO work, and related reports and confirmed our finding with stakeholders we interviewed, including those from USGCRP and the Mitigation Framework Leadership Group.

purposes. In addition, the National Climate Assessment provides high-level information on what is known about observed and projected climate risks in the United States. However, no federal entity looks holistically at the federal government's investments to strategically prioritize projects to ensure that they address the nation's most significant climate risks and provide the highest net benefits relative to other potential projects.

Further, we reported in September 2017 that the federal government had not undertaken strategic government-wide planning to manage significant climate risks before they become fiscal exposures.[46] As an initial step in managing climate risks, most of the experts we interviewed for the September 2017 report told us that federal decision makers should prioritize risk management efforts on significant climate risks that create the greatest fiscal exposure.[47]

Moreover, several stakeholders told us that the federal government's emphasis has been on funding postdisaster efforts instead of funding resilience projects before a disaster occurs. This is consistent with findings from our July 2015 report that most federal funding for hazard mitigation is only available after a disaster.[48] In addition, according to FEMA officials, some of the agency's hazard mitigation programs are designed to empower state and local governments to determine their mitigation funding priorities, and these state and local priorities may or may not align with the federal interest.

Finally, although we did not identify a government-wide strategic approach specifically for investing in climate resilience projects, the National Mitigation Investment Strategy—a national effort under way to plan for predisaster resilience investments—represents a potential cross-agency vehicle for climate resilience planning. However, the strategy does

[46] GAO-17-720.

[47] GAO-17-720.

[48] For example, we reported that for fiscal years 2011 through 2014, FEMA obligated more than \$3.2 billion for the Hazard Mitigation Grant Program for postdisaster hazard mitigation while obligating approximately \$222 million for the Pre-Disaster Mitigation Grant Program. GAO-15-515. In providing comments on a draft version of this chapter, FEMA officials told us that postdisaster hazard mitigation assistance could also reduce the impacts of future events.

not specifically address climate change or identify and prioritize specific climate resilience projects.

The Federal Government Could Reduce Its Fiscal Exposure by Focusing and Coordinating Federal Efforts

As we reported in March 2019, the federal government could reduce its fiscal exposure to climate change by focusing and coordinating federal efforts.[49] However, the federal government is currently not well organized to address the fiscal exposure presented by climate change, partly because of the inherently complicated and crosscutting nature of the issue. We have made a total of 62 recommendations related to limiting the federal government's fiscal exposure to climate change over the years, 12 of which have been made since February 2017.[50] As of December 2018, 25 of these recommendations remained open. In describing what needs to be done to reduce federal fiscal exposure to climate change, our March 2019 High-Risk Report discusses many of the open recommendations.[51] Implementing these recommendations could help reduce federal fiscal exposure. Several of them, including those highlighted below, identify key government-wide

---

[49] GAO-19-157SP.

[50] In June 2019, we reported that we had made a total of 62 recommendations related to limiting the federal government's fiscal exposure to climate change over the years, 12 of which had been made since February 2017. GAO, *Climate Change: Opportunities to Reduce Federal Fiscal Exposure*, GAO-19-625T (Washington, D.C.: June 11, 2019).

[51] We update our High-Risk List every 2 years. To determine which federal government programs and functions should be designated high risk, we consider qualitative factors, such as whether the risk could result in significantly impaired service or significantly reduced economy, efficiency, or effectiveness; the exposure to loss in monetary or other quantitative terms; and corrective measures planned or under way. We have issued the following five criteria for an area to be removed from the list: leadership commitment, capacity, action plan, monitoring, and demonstrated progress. In the March 2019 report, the federal government regressed in progress toward meeting the monitoring criterion for the *Limiting the Federal Government's Fiscal Exposure by Better Managing Climate Change Risks* high-risk area. Criteria for removing this area from the High-Risk List include demonstrating leadership commitment that is sustained and enhanced to address all aspects of the federal fiscal exposure to climate change cohesively. See GAO-19- 157SP.

efforts needed to help plan for and manage climate risks and direct federal efforts toward common goals, such as improving resilience.

- *Develop a national strategic plan:* In May 2011, we recommended that appropriate entities within the Executive Office of the President (EOP), including the Office of Management and Budget, work with agencies and interagency coordinating bodies to establish federal strategic climate change priorities that reflect the full range of climate- related federal activities, including roles and responsibilities of key federal entities.[52]
- *Use economic information to identify and respond to significant climate risks:* In September 2017, we recommended that the appropriate entities within EOP use information on the potential economic effects of climate change to help identify significant climate risks facing the federal government and craft appropriate federal responses.[53] Such federal responses could include establishing a strategy to identify, prioritize, and guide federal investments to enhance resilience against future disasters.
- *Provide decision makers with the best-available climate information:* In November 2015, we reported that federal efforts to provide information about climate change impacts did not fully meet the climate information needs of federal, state, local, and private sector decision makers, which hindered their efforts to plan for climate change risks.[54] We reported that these decision makers would benefit from a national climate information system that would develop and update authoritative climate observations and projections specifically for use in decision-making. As a result, we recommended that EOP (1) designate a federal entity to develop and periodically update a set of authoritative climate observations and projections for use in federal decision-making, which other

[52] EOP neither agreed nor disagreed with our recommendation and, as of March 2019, had not implemented it. GAO-11-317.
[53] EOP neither agreed nor disagreed with this recommendation and, as of March 2019, had not implemented it. GAO-17-720.
[54] GAO-16-37.

decision makers could also access, and (2)designate a federal entity to create a national climate information system with defined roles for federal agencies and nonfederal entities with existing statutory authority.[55]

- *Consider climate information in design standards:* In November 2016, we reported that design standards, building codes, and voluntary certifications established by standards-developing organizations play a role in ensuring the resilience of infrastructure to the effects of natural disasters.[56] However, we reported that these organizations faced challenges in using forward-looking climate information that could help enhance the resilience of infrastructure. As a result, we recommended in the November 2016 report that the Department of Commerce (Commerce), acting through the National Institute of Standards and Technology— which is responsible for coordinating federal participation in standards organizations— convene federal agencies for an ongoing government-wide effort to provide the best-available forward-looking climate information to standards-developing organizations for their consideration in the development of design standards, building codes, and voluntary certifications.[57]

In addition, in October 2019, we recommended that Congress consider establishing a federal organizational arrangement to periodically identify and prioritize climate resilience projects for federal investment.[58] We also identified six key steps the federal government could use to prioritize climate resilience investments and opportunities to increase the climate

---

[55] EOP neither agreed nor disagreed with these recommendations and, as of March 2019, had not implemented them.

[56] GAO, *Climate Change: Improved Federal Coordination Could Facilitate Use of Forward-Looking Climate Information in Design Standards, Building Codes, and Certifications*, GAO-17-3 (Washington, D.C.: Nov. 30, 2016).

[57] Commerce neither agreed nor disagreed with this recommendation and, as of May 2018, had not implemented it.

[58] GAO-20-127.

resilience impacts of federal funding options that Congress could use in designing the arrangement.[59]

In October 2019 we also issued the *Disaster Resilience Framework* to serve as a guide for analysis of federal action to facilitate and promote resilience to natural disasters.[60] The framework identifies three key principles that can help federal efforts to promote disaster resilience, including building resilience to climate change. First, authoritative and understandable information can help decision makers identify current and future risks and the impact of risk-reduction strategies. Second, integrated analysis and strategic planning can help decision makers take coherent and coordinated resilience actions. Third, financial and nonfinancial incentives can help make long-term, forward-looking risk-reduction investments more viable and attractive among competing priorities.

In conclusion, the effects of climate change have already posed and will continue to pose risks that can create fiscal exposure across the federal government, and this exposure will continue to increase. The federal government does not generally account for such fiscal exposure to programs in the budget process, and it has not undertaken strategic efforts to manage significant climate risks that could reduce the need for far more costly steps in the decades to come. To reduce its fiscal exposure, the federal government needs a cohesive strategic approach with strong leadership and the authority to manage risks across the entire range of related federal activities. The federal government could make further progress toward reducing fiscal exposure by implementing the recommendations we have made.

Chairman Rouda, Ranking Member Comer, and Members of the Subcommittee, this completes my prepared statement. I would be pleased to respond to any questions that you may have at this time.

---

[59] The six key steps we identified were (1) define the effort's strategical goals and provide an entity with authority to lead the effort;(2) identify and assess high-risk regions, sectors, and climate risks; (3) identify potential project ideas by seeking proposals or via an expert panel; (4) prioritize projects using key criteria; (5) implement high-priority projects when funds are available; and (6) monitor projects and climate risks to inform future decisions.

[60] GAO, Disaster Resilience Framework: Principles for Analyzing Federal Efforts to Facilitate and Promote Resilience to Natural Disasters, GAO-20-100SP (Washington, D.C.: October 2019).

In: Climate Change
Editor: Iyana Hendricks
ISBN: 978-1-53617-757-2

Chapter 4

ATTACHING A PRICE TO GREENHOUSE GAS EMISSIONS WITH A CARBON TAX OR EMISSIONS FEE: CONSIDERATIONS AND POTENTIAL IMPACTS*

Jonathan L. Ramseur and Jane A. Leggett

ABSTRACT

The U.S. Fourth National Climate Assessment, released in 2018, concluded that "the impacts of global climate change are already being felt in the United States and are projected to intensify in the future—but the severity of future impacts will depend largely on actions taken to reduce greenhouse gas [GHG] emissions and to adapt to the changes that will occur." Members of Congress and stakeholders articulate a wide range of perspectives over what to do, if anything, about GHG emissions, future climate change, and related impacts. If Congress were to consider establishing a program to reduce GHG emissions, one option would be to attach a price to GHG emissions with a carbon tax or GHG emissions fee.

* This is an edited, reformatted and augmented version of Congressional Research Service, Publication No. R45625, dated March 22, 2019.

In the 115th Congress, Members introduced nine bills to establish a carbon tax or emissions fee program. However, many Members have expressed their opposition to such an approach. In particular, in the 115th Congress, the House passed a resolution "expressing the sense of Congress that a carbon tax would be detrimental to the United States economy."

Multiple economic studies have estimated the emission reductions that particular carbon tax would achieve. For example, a 2018 study analyzed various impacts of four carbon tax rate scenarios: a \$25/metric ton of CO_2 and \$50/metric ton of CO_2 carbon tax, increasing annually by 1% and 5%. The study concluded that each of the scenarios would likely achieve the U.S. GHG emission reduction target pledged under the international Paris Agreement (at least in terms of CO2 emissions).

A carbon tax system would generate a new revenue stream, the magnitude of which would depend on the scope and rate of the tax, among other factors. In 2018, the Congressional Budget Office (CBO) estimated that a \$25/metric ton carbon tax would yield approximately \$100 billion in its first year. CBO projected that federal revenue would total \$3.5 trillion in FY2019.

Policymakers would face challenging decisions regarding the distribution of the new carbon tax revenues. Congress could apply revenues to support a range of policy objectives but would encounter trade-offs among the objectives. The central trade-offs involve minimizing economy-wide costs, lessening the costs borne by specific groups—particularly low-income households and displaced workers in certain industries (e.g., coal mining)—and supporting other policy objectives.

A primary argument against a carbon tax regards it potential economy-wide impacts, often measured as impacts to the U.S. gross domestic product (GDP). Some may argue that projected impacts should be compared with the climate benefits achieved from the program as well as the estimated costs of taking no action. The potential impacts would depend on a number of factors, including the program's magnitude and design and, most importantly, the use of carbon tax revenues.

In general, economic literature finds that some of the revenue applications would reduce the economy-wide costs from a carbon tax but may not eliminate them entirely. In addition, some studies cite particular economic modeling scenarios in which certain carbon tax revenue applications produce a net increase in GDP compared to a baseline scenario. These scenarios involve using carbon tax revenues to offset reductions in other tax rates (e.g., corporate income or payroll taxes). Although economic models generally indicate that these particular revenue applications would yield the greatest benefit to the economy overall, the models also find that lower-income households would likely face a disproportionate impact under such an approach. As lower-income households spend a greater proportion of their income on energy needs (electricity, gasoline), these households are expected to experience disproportionate impacts from a carbon tax if revenues were not recycled back to them in some fashion (e.g., lump-sum distribution).

A price on GHG emissions could create a competitive disadvantage for some industries, particularly "emission-intensive, trade-exposed industries." Policymakers have several options to address this concern, including establishing a "border carbon adjustment" program, which would levy a fee on imports from countries without comparable GHG reduction programs. Alternatively, policymakers could allocate (indefinitely or for a period of time) some of the carbon tax revenues to selected industry sectors or businesses. Relatedly, a carbon tax system is projected to disproportionately impact fossil fuel industries, particularly coal, and the communities that rely on their employment. To alleviate these impacts, policymakers may consider using some of the revenue to provide transition assistance to employees or affected communities.

INTRODUCTION

The U.S. Fourth National Climate Assessment, released in 2018, concluded that "the impacts of global climate change are already being felt in the United States and are projected to intensify in the future—but the severity of future impacts will depend largely on actions taken to reduce greenhouse gas [GHG] emissions and to adapt to the changes that will occur."[1] Although a variety of efforts seeking to reduce GHG emissions are currently underway on the international[2] and sub-national level,[3]

---

[1] U.S. Global Change Research Group, Fourth National Climate Assessment Volume II: Impacts, Risks, and Adaptation in the United States, Chapter 1, 2018, https://nca2018.globalchange.gov.

[2] For more details, see Organization for Economic Cooperation and Development, *Effective Carbon Rates 2018: Pricing Carbon Emissions Through Taxes and Emissions Trading*, 2018, http://www.oecd.org/tax/effective-carbonrates-2018-9789264305304-en.htm; and the Carbon Tax Center website at http://www.carbontax.org/where-carbon-istaxed.

[3] A number of U.S. states have taken action requiring GHG emission reductions, including California and the Regional Greenhouse Gas Initiative (RGGI)—a coalition of nine states from the Northeast and Mid-Atlantic regions. The RGGI is a cap-and-trade system that took effect in 2009 and applies to carbon dioxide emissions from electric power plants. (See CRS Report R41836, *The Regional Greenhouse Gas Initiative: Lessons Learned and Issues for Congress*, by Jonathan L. Ramseur.) California established a cap-and-trade program that took effect in 2013. California's cap covers multiple GHGs, which account for

federal policymakers and stakeholders have different viewpoints over what to do, if anything, about future climate change and related impacts. Their views regarding climate change cover a wide range of perspectives.[4]

For example, some contend that climate change poses a "direct, existential threat"[5] to human society and that nations must start making significant reductions in GHG emissions in order to avoid "dire effects."[6] To support this argument, proponents of climate change mitigation highlight the evidence and conclusions from recent reports that are generally considered authoritative, including:

1. The Intergovernmental Panel on Climate Change, *Global Warming of 1.5°C*, 2018;[7] and
2. The U.S. Global Change Research Program, *Fourth National Climate Assessment, Volume II: Impacts, Risks, and Adaptation in the United States*, 2018.[8]

On the other hand, some question whether there are sufficient risks of climate change to merit a federal program requiring GHG emission

approximately 85% of California's GHG emissions. For more details, see the California Air Resources Board website, https://www.arb.ca.gov/cc/capandtrade/capandtrade.htm.

[4] An exhaustive documentation of these varied perspectives is beyond the scope of this chapter.

[5] Speech by United Nations Secretary-General Antonio Guterres, September 10, 2018, https://www.un.org/sg/en/ content/sg/statement/2018-09-10/secretary-generals-remarks-climate-change-delivered.

[6] See, for example, Heather Smith, "Climate Change: Even Worse That We Thought," *Sierra Club Magazine*, October 2018.

[7] Intergovernmental Panel on Climate Change (IPCC), *Global Warming of 1.5°C*, 2018, http://www.ipcc.ch/report/sr15/. The IPCC is organized under the auspices of the United Nations. According to the IPCC website, "thousands of people from all over the world contribute to the work of the IPCC.... An open and transparent review by experts and governments around the world is an essential part of the IPCC process, to ensure an objective and complete assessment and to reflect a diverse range of views and expertise. Through its assessments, the IPCC identifies the strength of scientific agreement in different areas and indicates where further research is needed. The IPCC does not conduct its own research" (https://www.ipcc.ch/about/).

[8] Established by the Global Change Research Act of 1990 (P.L. 101-606), the U.S. Global Change Research Program coordinates and integrates federal research and applications to understand, assess, predict, and respond to human-induced and natural processes of global change. Thirteen federal agencies and departments participate in the program.

reductions.[9] In addition, others argue that a unilateral approach to climate change by the United States could disproportionately impact domestic industries while achieving minimal results in global climate change mitigation.[10]

If Congress were to consider establishing a program to reduce GHG emissions, one option would be to apply a tax or fee on GHG emissions or the inputs that produce them. This type of approach is commonly called a carbon tax or a GHG emissions fee (see "Terminology Issues: A Carbon Tax or an Emissions Fee?").

Terminology Issues: A Carbon Tax or an Emissions Fee?

In the context of carbon price policy, terminology may be a key issue. As many policymakers, stakeholders, and academic journals use the term *carbon tax*, this is the default term in this chapter. Related terms cited in economic literature include *emissions fee* or *emissions charge*. Several proposals in recent Congresses described their approach as a GHG emissions fee. Whether the policy proposal is labeled as a tax, fee, or other term, the carbon price may apply only to carbon dioxide (CO_2) emissions or multiple GHG emissions.

The choice of terminology between a tax and fee may have procedural consequences, particularly in terms of congressional committee jurisdiction. For instance, a carbon tax proposal may involve a referral to the House Committee on Ways and Means (in addition to any other House committees of jurisdiction). In the Senate, tax measures are referred to the Committee on Finance. Standing committees of the House, other than the Appropriations and Budget Committees, may report legislation creating or modifying user fees.[11] Committees with jurisdiction over environmental policy (e.g., House Committee on Energy and Commerce and Senate Committee on Environment and Public Works) may have different perspectives, expertise, or policy objectives than tax committees. These differences could potentially influence the design of a carbon price instrument. In addition, there may be legal considerations depending on whether the program is structured as a fee or tax. These issues are beyond the scope of this chapter.

---

9 See, for example, David Kreutzer el al., "The State of Climate Science: No Justification for Extreme Policies," Heritage Foundation, 2016, https://www.heritage.org/environment/report/the-state-climate-science-no-justificationextreme-policies; Marlo Lewis, "You Don't Have to Be a Climate Skeptic to Oppose a Carbon Tax," Competitive Enterprise Institute, 2018, https://cei.org/blog/you-dont-have-be-climate-skeptic-oppose-carbon-tax.

10 See, for example, letter from 20 organizations to Paul Ryan (then-Speaker of the House) and Kevin McCarthy (then-Majority Leader), July 9, 2018, https://www.americanenergyalliance.org/wp-content/uploads/2018/07/ CarbonTaxLetterUpdated.pdf; and Oren Cass, "The Carbon-Tax Shell Game," *National Affairs*, 2015.

11 However, this procedural point would likely be the subject of debate: A ruling on whether a policy instrument is a user fee or a tax measure may depend on the nature of the charge rather than its label. This issue is beyond the scope of this chapter.

This chapter does not compare and analyze the multiple policy tools available to Congress that could address climate change (see text box "Other Policy Options for Addressing GHG Emissions"). This chapter focuses on the policy considerations and potential impacts of using a carbon tax or GHG emissions fee to control GHG emissions.

The key human-related GHG is CO_2, which is generated primarily through the combustion of fossil fuels: coal, oil, and natural gas. In 2016, fossil fuel combustion accounted for 94% of U.S. CO_2 emissions and 76% of U.S. GHG emissions.[12] A carbon tax could apply either directly to GHG emissions or to the materials—based on their carbon contents—that ultimately generate the emissions (i.e., "emissions inputs").[13] A carbon price on emissions or their emissions inputs—mainly fossil fuels—would increase the relative price of the more carbon-intensive energy sources,[14] particularly coal. This result could spur innovation in less carbon-intensive technologies (e.g., renewable energy, nuclear power, carbon capture and sequestration [CCS]) and stimulate other behavior that may decrease emissions, such as efficiency improvements. The energy price increases could also have both economy-wide impacts and negative effects on specific industries and particular demographic groups.

A carbon tax approach has received some attention and debate in recent years. In the 115th Congress, Members introduced nine carbon tax or fee proposals.[15] Outside of Congress, the Climate Leadership Council—a bipartisan group of former policymakers and industry leaders— published

---

[12] As expressed in CO2-equivalents. See U.S. Environmental Protection Agency, *Inventory of U.S. Greenhouse Gas Emissions and Sinks*, 1990-2016, April 2018.

[13] This differs from a price system that applies to energy content, such a tax based on British thermal units (Btu). In 1993, President Clinton proposed a deficit reduction package that included a tax based on energy content, measured in Btu. The goals of the proposal were to promote energy conservation and raise revenue. At the time, the proposed tax would have generated a new revenue stream of about $30 billion per year. The proposal was met with strong opposition and was not enacted. Congress ultimately enacted an approximately 5-cent per gallon increase in the motor fuels taxes.

[14] Carbon intensity refers to the amount of CO2 emissions generated per unit of energy. The CO2 emission intensity of coal is approximately 30% more than oil and approximately 80% more than natural gas.

[15] For more details, see CRS Report R45472, *Market-Based Greenhouse Gas Emission Reduction Legislation: 108th Through 115th Congresses*, by Jonathan L. Ramseur. Other legislation may indirectly reduce GHG emissions, for example, by providing tax or other incentives for renewable energy or CCS activities.

a conceptual carbon tax approach in 2017 that generated some interest.[16] Some of the industry leaders on the council represent major energy companies, including Shell, BP, and ExxonMobil.

Other Policy Options for Addressing GHG Emissions

For policymakers considering actions to address climate change, a variety of policy instruments are available. Another option for directly reducing GHG emissions is to impose an emissions cap, likely complemented with an emissions trading program (i.e., cap-and-trade). To some extent, a carbon tax and cap-and-trade program would produce similar effects: Both would place a price on carbon, both could increase the price of fossil fuels, and both could reduce GHG emissions. Preference between the two approaches ultimately depends on which variable policymakers prefer to more precisely control—emission levels or emission prices. As a practical matter, these market-based policies may include complementary or hybrid designs, incorporating elements to increase price certainty or emissions quantity certainty. For example, legislation could provide mechanisms for adjusting a carbon tax if a targeted range of emissions reductions were not achieved in a given period. Alternatively, legislation could include mechanisms that would bound the range of market prices for emissions allowances to improve price certainty.

Although recent attention has largely focused on market-based mechanisms such as a carbon tax and cap-and-trade programs, nonmarket policy tools may be an option to address some emission sources. For example, Congress has already addressed emissions from cars, light trucks, and government buildings through performance standards. In addition, Congress may continue to support the development of GHG emission mitigation technologies, such as CCS, or as a supplement to the primary climate change mitigation policy.[17]

On the other hand, many Members have expressed their opposition to a carbon tax. Starting in the 112th Congress and going through the 115th Congress, Members have introduced resolutions in both the House and Senate expressing the view that a carbon tax is not in the economic interests of the United States. In 2018, the House passed a resolution "expressing the sense of Congress that a carbon tax would be detrimental to the United States economy" (H.Con.Res. 119).[18] An analogous resolution was not introduced in the Senate in the 115th Congress.

The first section of this chapter examines carbon tax design issues, including the point of taxation, the rate of taxation, and potential border carbon adjustments. The second section discusses issues related to the

[16] For more details, see the council's website at https://www.clcouncil.org/.

[17] For more information, see CRS In Focus IF10589, *FY2019 Funding for CCS and Other DOE Fossil Energy R&D*, by Peter Folger.

[18] The House passed an identical resolution in the 114th Congress (H.Con.Res. 89).

distribution of carbon tax revenues. The third section discusses additional considerations associated with a carbon tax program, including estimates of GHG emissions, federal revenue, and fossil fuel prices and changes in energy use. The fourth section provides concluding observations.

CARBON TAX DESIGN CONSIDERATIONS

If policymakers decide to establish a carbon tax system, Congress would face several key design decisions, including the point of taxation—where to impose the tax and what to tax—the rate of taxation, and whether and/or how to address imported carbon-intensive materials.

Alternatively, Congress could direct one or more federal agencies to determine these design features through a rulemaking procedure. Although a few of the GHG emission reduction proposals in prior Congresses delegated such authority to an agency, such as the U.S. Environmental Protection Agency (EPA),[19] all of the proposals since the 111th Congress have included some degree of design details in the statutory language. (A later section discusses carbon tax revenue application considerations).

Point of Taxation

The point of taxation would determine which entities would be required to (1) make tax payments based on emissions or emission inputs, such as fossil fuels, (2) monitor emissions or emission inputs, and (3) maintain records of relevant activities and transactions. This section provides some considerations for policymakers deciding which GHG emissions and/or emission sources to cover in a carbon tax system.

Throughout the U.S. economy, millions of discrete sources generate GHG emissions: power plants, industrial facilities, motor vehicles,

---

[19] See, for example, H.R. 2042 in the 108th Congress, S. 3639 in the 109th Congress, and S. 309 in the 110th Congress.

households, commercial buildings, livestock, etc. Administrative costs and challenges would likely increase with a broader scope of an emissions tax.

A carbon tax may apply to CO_2 emissions alone, which account for most U.S. GHG emissions, or to multiple GHGs. Carbon tax proposals that apply only to CO_2 generally attach a price to a metric ton of CO_2 emissions ($mtCO_2$). Some sources emit non-CO_2 GHGs, such as methane, nitrous oxides, and sulfur hexafluoride. GHG emissions from these sources could be addressed by attaching a price to a metric ton of CO_2 emissions-equivalent ($mtCO_2e$). This term of measure is used because GHGs vary by global warming potential (GWP). At these sources, the determined GWP values would be an important issue.[20]

Policymakers may consider limiting the tax to sectors or sources that emit above a certain percentage of total U.S. GHG emissions, many of which currently report their emissions to the government. For more than 20 years, monitoring devices or systems have been installed in smokestacks of most large facilities, such as power plants, which are required to periodically report emissions data to EPA.[21] In addition, since 2010, EPA has collected annual emissions data from approximately 8,000 facilities that directly release above certain amounts of GHG emissions.[22] Using these established monitoring frameworks, policymakers could employ a "downstream" approach, applying a carbon tax at the point where the GHGs from these facilities are released to the atmosphere.

---

[20] GWP is an index developed by the IPCC that allows comparisons of the heat-trapping ability of different gases over a period of time, typically 100 years. Consistent with international GHG reporting requirements, EPA's most recent GHG inventory (2018) uses the GWP values presented in the IPCC's 2007 Fourth Assessment Report. For example, based on these GWP values, a metric ton of methane equates to 25 metric tons of CO2 when averaged over a 100-year time frame. The IPCC has since updated the 100-year GWP estimates, with some increasing and some decreasing. For example, the IPCC 2013 Fifth Assessment Report reported the 100-year GWP for methane as ranging from 28 to 36. The uncertainty in the GWP for a particular GHG could be of interest for policymakers. See Donald Marron et al., *Taxing Carbon: What, Why, and How*, Tax Policy Center, 2015, p. 4.

[21] 40 C.F.R. Part 75.

[22] The FY2008 Consolidated Appropriations Act (P.L. 110-161) directed EPA to establish a mandatory reporting program for "appropriate thresholds in all sectors of the economy." EPA established its Greenhouse Gas Reporting Program in 2009, which is codified in 40 C.F.R. Part 98. For more information, see http://www.epa.gov/ghgreporting.

Alternatively, the tax could be applied to reliable proxies for emissions, such as emission inputs. For example, the carbon content of fossil fuels—coal, natural gas, petroleum—can serve as a proxy for the emissions released when the fuels are combusted.[23] Applying a tax on emission inputs allows for the consideration of various points of taxation. For instance, emission inputs could be taxed at "upstream" (e.g., wells) or "midstream" stages in that process (e.g., refineries), the latter allowing for potential tax administration advantages that may be provided by specific infrastructure chokepoints in the fossil fuel market. For example, with respect to petroleum, the number of upstream sources—wells that produce crude oil—is over 445,000, but the number of midstream sources—facilities that refine crude oil—is only 137.

Table A.1 (in Appendix A) lists the top GHG emission sources in the United States. These sources combined to account for approximately 95% of U.S. GHG emissions in 2016. Table A.1 identifies the number of entities for each source category (e.g., number of coal mines, number of steel production facilities) and the percentage of total U.S. GHG emissions the category contributes.

In the case of fossil fuel combustion—which accounted for 76% of total U.S. GHG emissions— the table provides several options for segmenting the universe of sources if policymakers choose to implement a carbon tax. It identifies the number of entities that might be subject to the carbon tax under a particular option (pending any exclusions). For example, policymakers could address fossil fuel combustion emissions by applying a carbon tax to fossil fuels (based on their carbon content) at the following entities, which include both upstream and midstream infrastructure chokepoints:

- 137 petroleum refineries (based on 2017 data) and 166 petroleum importers (based on 2018 data);
- 671 coal mines and eight companies supplying imported coal (based on 2017 data); and

[23] EPA's GHG reporting regulations also apply to suppliers of fossil fuels, which include nearly 1,000 fossil-fuelreporting suppliers.

- 1,679 entities that report natural gas deliveries to the Energy Information Administration (EIA) on Form EIA-176 and 123 natural gas fractionators[24] (based on 2016 data).

Some of the above points of taxation might take advantage of the administrative frameworks for existing federal excise taxes. For example, a per-barrel federal excise tax on crude oil at the refinery supports the Oil Spill Liability Trust Fund.[25] An excise tax on the sale or use of coal supports the Black Lung Disability Trust Fund.[26]

Rate of Carbon Tax

A central policy choice when establishing a price on GHG emissions is the rate of the carbon tax. Several approaches, which are discussed below, could inform the decision.

GHG Emissions Target Approach

One approach would set the carbon tax rate at a level or pathway—based on modeling estimates—that would achieve a specific GHG emissions target. For example, a 2018 study estimated the carbon tax rate needed to meet the U.S. GHG emission reduction targets established under the 2015 Paris Agreement: 26%-28% below 2005 net GHG emission levels by 2025.[27] The study found that a constant tax rate of $43/ton starting in 2019 would meet the 2025 reduction target.[28]

---

[24] A natural gas fractionator separates natural gas liquids (ethane, propane, butane) from the natural gas stream.

[25] See CRS In Focus IF10823, *The Oil Spill Liability Trust Fund Tax: Reauthorization Issues and Legislation in the 115th Congress*, by Jonathan L. Ramseur.

[26] See CRS Report R45261, *The Black Lung Program, the Black Lung Disability Trust Fund, and the Excise Tax on Coal: Background and Policy Options*, by Scott D. Szymendera and Molly F. Sherlock.

[27] For a description of the current U.S. commitment, see CRS In Focus IF10239, *President Obama Pledges Greenhouse Gas Reduction Targets as Contribution to 2015 Global Climate Change Deal*, by Jane A. Leggett. President Trump announced an intention to withdraw the United States from the agreement when eligible in 2020. No action has been taken to withdraw or modify the U.S. pledged GHG reduction contribution. See CRS In

Emissions reduction estimates from carbon tax programs are based on multiple assumptions. Accordingly, such estimates provide different tax rates needed to meet a particular emissions target depending on these assumptions.[29] See "Impacts on GHG Emission Levels" for selected analyses of emission reductions for a given carbon price and rate of price increase.

Marginal Benefits or "Social Cost of Carbon" Approach

Under another approach, policymakers could base the carbon tax rate on the estimated marginal[30] net benefits associated with reduced CO_2 emissions.[31] The net benefits would be the avoided net damages (i.e., costs) of climate change.[32] The estimates of net benefits of avoided emissions often rely on analyses of the *social cost of carbon (SC-CO_2)* or the *social cost of greenhouse gases (SCGHG)*.[33] Therefore, policymakers could use SC-CO_2 measurements—as the basis for an estimate of the net benefits of a

Focus IF10668, *Potential Implications of U.S. Withdrawal from the Paris Agreement on Climate Change*, by Jane A. Leggett.

[28] Yunguang Chen and Marc Hafstead, "Using a Carbon Tax to Meet U.S. International Climate Pledges," *Climate Change Economics*, 2018. The modeled carbon tax system applied to emissions from fossil fuel combustion.

[29] For example, multiple parties prepared estimates during the development of H.R. 2454 (which passed the House on June 26, 2009) in the 111th Congress. That proposal intended to reduce GHG emissions to 17% below 2005 levels by 2020. The emission allowance price estimates needed to meet emission targets ranged from $16/mtCO2e to $49/mtCO2e in 2015, with the estimated range increasing over time.

[30] By "marginal," economists mean an additional, small increment—in other words, the benefits associated with making an additional, small amount of emissions reduction, not the total emissions reductions from a policy.

[31] See, for example, Organisation for Economic Co-operation and Development, "The Social Cost of Carbon," in *Cost-Benefit Analysis and the Environment: Further Developments and Policy Use*, 2018.

[32] Net benefits refers to the beneficial value of abating GHG emissions (e.g., avoiding heat stress deaths) *minus* the positive impacts that climate change could bring (e.g., increased agricultural productivity in some regions). One challenge in these metrics is that the damages and the benefits of climate change often impact different groups and sometimes at different times. In addition, the calculation of costs and benefits can vary based on the geographic scope considered. See CRS Report R45119, *EPA's Proposal to Repeal the Clean Power Plan: Benefits and Costs*, by Kate C. Shouse.

[33] For more information, see CRS In Focus IF10625, *Social Costs of Carbon/Greenhouse Gases: Issues for Congress*, by Jane A. Leggett.

marginal change in emissions—to set the rate of a carbon tax or emissions fee.[34]

One potential challenge of relying on SC-CO_2 estimates to set a carbon fee are methodological concerns.[35] For example, the existing estimates in peer-reviewed research cover a wide range. In addition, some argue that the underlying simulation models for estimating the SC-CO_2 values are insufficient. For any level of emissions, the projected increase in global average temperature may cover multiple degrees Fahrenheit, and other measures of climate change, such as precipitation patterns, may encompass directional uncertainties. No estimates of impacts are comprehensive at this time, and many of the risks are difficult to estimate and value.

When valuing the SC-CO_2, analysts encounter a range of views on methods and assumptions, and establishing study parameters may be challenging. For example, estimates of the monetary values of climate change impacts may be difficult or controversial to estimate, such as the monetary values associated with human deaths or sickness. A related framework question is whether to include global climate impacts or just domestic impacts.[36]

In addition, the element of time in climate change impacts particularly complicates the valuation. The fact that many impacts of climate change will occur in the distant future requires consideration of society's willingness to pay *in the near term* to reduce emissions that would cause *future* damages, mostly to future generations. To take time into account,

---

[34] The SC-CO2 measures the net benefits of relatively small (marginal) reductions in CO2 emissions. The federal SCCO2 values used in rulemakings have been based on business-as-usual projections, which may not reflect the economically efficient GHG emission levels in the long-term. See for example, International Monetary Fund, *Fiscal Policy to Mitigate Climate Change: A Guide for Policymakers* (Washington, DC: International Monetary Fund, 2012), https://www.imf.org/en/Publications/Books/Issues/2016/12/31/Fiscal-Policy-to-Mitigate-Climate-Change-A-Guide-forPolicymakers-25864.

[35] See public comments on the SC-CO2 for certain regulatory uses contained in the docket for Office of Management and Budget, *Technical Support Documents: Update of the Social Cost of Carbon for Regulatory Impact Analysis under Executive Order No. 12866*, 2014, https://www.regulations.gov/document?D=OMB-2013-0007-0063. See also Richard S. J. Tol, "Is the Uncertainty About Climate Change Too Large for Expected Cost-Benefit Analysis?," *Climatic Change*, 2003.

[36] See the section "Considerations for the Scope—Domestic or Global—of the SCC [i.e., SC-CO2]" in CRS Report R45119, *EPA's Proposal to Repeal the Clean Power Plan: Benefits and Costs*, by Kate C. Shouse.

economists discount future values to a calculated "present value." Economists do not agree on the appropriate discount rate(s) to use for a multi-generational, largely nonmarket issue such as human-induced climate change. The choice of discount rate can significantly increase or decrease values of the SC-CO_2. A low discount rate would give greater value today to future impacts than would a higher discount rate. High discount rates can reduce the value today of future climate change impacts to a small fraction of their undiscounted values. A high discount rate would recommend applying fewer of today's resources to addressing climate change impacts in the future.

Since 2008, federal agencies have used SC-CO_2 estimates in dozens of final rulemakings as a method to estimate the net benefits of abating CO_2 emissions.[37] An Interagency Working Group prepared SC-CO_2 estimates, which were updated over time and subjected to expert and public comment. On March 28, 2017, President Trump issued Executive Order 13783, "Promoting Energy Independence and Economic Growth," which effectively withdrew the federal SC-CO_2 estimates. Nonetheless, federal agencies have used new, interim values generated by EPA in 2017, modified from the withdrawn technical support documents, in regulatory and other decisions. Legislation could set a carbon price citing any of these SC-CO_2 values or others available from nonfederal researchers or prescribe methods for estimating new ones.

Using SC-CO_2 estimates to set the tax rate would involve a cost-benefit framework. Although many posit that a cost-benefit framework remains the best option,[38] some economists argue that a cost-benefit framework may be inappropriate for climate change policy for these reasons[39]

---

[37] See CRS Report R44657, *Federal Citations to the Social Cost of Greenhouse Gases*, by Jane A. Leggett.

[38] Policy analysts have been exploring options to mitigate some of the problems in applying this approach to climate change. See, for example, Francis Dennig, "Climate Change and the Re-Evaluation of Cost-Benefit Analysis," *Climatic Change*, 2017.

[39] See, for example, M. Granger Morgan et al. "Why Conventional Tools for Policy Analysis Are Often Inadequate for Problems of Global Change," *Climatic Change*, vol. 41 (1999), pp. 271-281; Robert S. Pindyck, "Uncertainty in Environmental Economics," in *Review of Environmental Economics and Policy* (Oxford: Oxford University Press, 2007).

- Many experts expect climate change—and policies to address it—to cause nonmarginal changes to economies and ecosystems. The changes are expected to increase disproportionately with incremental climate change with a potential for crossing critical "tipping points" after which systems change dramatically and rapidly.[40]
- Climate change impacts are multi-generational, and uncertainty and disagreement exists about whether and how to assign a present value to social costs and benefits over generations.
- Some impacts from climate change may be irreversible on the timescale of human civilizations, such as melting of major ice sheets in Antarctica or Greenland.[41]

Other Considerations

Policymakers might consider a carbon tax as a fiscal tool to help reduce the federal deficit, reduce other taxes, or pay for specific programs that may or may not be related to climate change policy. In addition, some have proposed a phased-in approach, setting a rate that is initially lower but increases at an announced or adjustable rate either for a fixed period or indefinitely. Advantages of this approach include providing an opportunity for consumers and investors to adjust their behavior before the higher tax rates go into effect, such as purchasing more energy efficient appliances or investing in low-emissions technologies. Phasing in a carbon tax, however, could delay climate-related benefits.

If Congress finds agreement in principle on carbon pricing, the rate(s) could emerge from the process of reaching political agreement. Elements that might be considered include the options described above or consideration of the magnitude of overall economic impact; impacts on certain economic sectors, regions, or population groups; timing to motivate and allow an orderly transition to a lower-GHG economy; or other factors.

---

[40] Examples of tipping points include "the destabilization of the West Antarctic ice sheet or rapid methane release from thawing permafrost." See the U.S. Global Change Research Program, *Fourth National Climate Assessment, Volume II: Impacts, Risks, and Adaptation in the United States*, 2018.

[41] See, for example, IPCC, *Global Warming of 1.5°C*, Chapter 3, http://www.ipcc.ch/report/sr15/.

Border Carbon Adjustments

Many stakeholders have voiced concerns over how a U.S. carbon price system would interact with policies in other nations, particularly if the United States were to enact a carbon tax system that covers more sources or is more stringent than enacted elsewhere. A central concern is that a U.S. carbon tax could raise U.S. prices more than the prices of goods manufactured abroad, potentially creating a competitive disadvantage for some domestic businesses.[42] Certain businesses may become less profitable, lose market share, and reduce jobs.

The industries generally expected to experience disproportionate impacts under a U.S. carbon tax are often described as "emission-intensive, trade-exposed" industries. An industry's CO_2 emission intensity is a function of both direct CO_2 emissions from its manufacturing process (e.g., CO_2 from cement or steel production) and indirect CO_2 emissions from the inputs to the manufacturing process (e.g., electricity, natural gas). Such industries are likely to experience greater cost increases than less carbon intensive industries, all else being equal. In general, trade-exposed industries are those that face greater international competition compared to other domestic industries. A carbon tax could present a particular challenge for these industries, because they might be less able to pass along the tax in the form of higher prices, because they may lose global market share—and jobs—to competitors in countries lacking comparable carbon policies.

Policymakers might consider approaches to mitigate these potential economic impacts in several ways. One approach that has received interest in recent years is a border adjustment mechanism, which is often described as a border carbon adjustment (BCA) in the carbon tax context. A BCA would apply a tariff to emission-intensive, imported goods such as steel, aluminum, cement, and certain chemicals. Each of the carbon price

[42] Not all businesses within a sector may be affected similarly. For example, under a carbon tax system, an aluminum company using electricity produced with hydropower would experience less cost increase than a company using electricity produced with coal. In addition, some businesses may be more energy efficient than others or use less emitting processes. Some may be able to reduce their emissions in response to a carbon tax at lower cost than others.

proposals in the 115th Congress would have established a BCA to address emission-intensive imports.[43]

Another rationale for adding a BCA to a carbon tax system is the possibility that it would encourage other nations to adopt comparable carbon price policies.[44] Many of the recently proposed BCA mechanisms allow for exemptions for nations with comparable programs.

To date, no nations have implemented a BCA as part of their climate change policies. Establishing an economically efficient BCA would likely present substantial challenges.[45] For example, policymakers must decide which goods and/or industries would be covered by a BCA and how the adjustment program would assess the comparability of varied climate-related policies in other nations. In addition, accurately determining and verifying the volume of GHG emissions embodied in a particular imported product would be data intensive and challenging. To alleviate some of the measurement complexity, policymakers could limit the program to selected industries and apply default values and assumptions to particular manufacturing processes. However, this simplified approach could result in less accurate import price adjustments, which could potentially affect the accuracy of GHG emission reductions achieved by the carbon tax program.[46] Another option would be to allow companies to provide measured, independently verified emissions data as an alternative to default values.

---

[43] See CRS Report R45472, *Market-Based Greenhouse Gas Emission Reduction Legislation: 108th Through 115th Congresses*, by Jonathan L. Ramseur.

[44] See, for example, Adele Morris, *Making Border Carbon Adjustments Work in Law and Practice*, Tax Policy Center, 2018.

[45] See, for example, Marco Sakai and John Barrett, "Border Carbon Adjustments: Addressing Emissions Embodied in Trade," *Energy Policy*, 2016; Sam Kortum and David Weisbach, "Border Adjustment for Carbon Emissions," Resources for the Future, 2016; Carolyn Fischer et al., "Carbon Taxes and Energy-Intensive Trade-Exposed Industries," in *Implementing a U.S. Carbon Tax: Challenges and Debates*, ed. Ian Parry et al. (Washington, DC: International Monetary Fund, 2015).

[46] Congressional Budget Office, *Border Adjustments for Economywide Policies That Impose a Price on Greenhouse Gas Emissions*, 2013.

In addition, the border adjustment approach would likely raise concerns of violating international trade rules.[47] Further, some researchers have highlighted the potential for unintended consequences from a BCA. For example, some studies have found that a border adjustment may lead to lower net exports than the carbon price alone, due to the adjustment's terms-of-trade effect on U.S. currency.[48] These issues are beyond the scope of this chapter, but some of the concerns may be lessened to some degree if a larger number of nations establish comparable emission reduction policies, as many have agreed to do under the Paris Agreement.[49]

Another possible rationale for a BCA is to address the concern of "emissions leakage" (or "carbon leakage"). Emissions leakage "occurs when economic activity is shifted as a result of the emission control regulation [e.g., a carbon tax program] and, as a result, emission abatement achieved in one location that is subject to emission control regulation is [diminished] by increased emissions in unregulated locations."[50] The concern of emissions leakage has been central in the debate over whether the United States (or any nation) should unilaterally address GHG emissions. A BCA may diminish the potential for emissions leakage by reducing the incentive to shift economic activity to a nation without a comparable carbon tax. However, some recent studies raise questions regarding the degree to which emissions leakage would be a concern under a unilateral U.S. carbon tax.[51]

---

[47] See, for example, Joel Trachtman, "WTO Law Constraints on Border Tax Adjustment and Tax Credit Mechanisms to Reduce the Competitive Effects of Carbon Taxes," Resources for the Future, 2016.

[48] See, for example, Warwick McKibbin et al., "The Role of Border Carbon Adjustments in a U.S. Carbon Tax," *Climate Change Economics*, vol. 9, no. 1 (2018).

[49] CRS Report R44609, *Climate Change: Frequently Asked Questions About the 2015 Paris Agreement*, by Jane A. Leggett and Richard K. Lattanzio.

[50] EPA, Office of Air and Radiation, *Tools of the Trade: A Guide to Designing and Operating a Cap and Trade Program for Pollution Control*, 2003, Glossary.

[51] See, for example, McKibbin et al., "The Role of Border Carbon Adjustments in a U.S. Carbon Tax"; Morris, *Making Border Carbon Adjustments Work in Law and Practice*, 2018; Joseph Aldy, "Frameworks for Evaluating Policy Approaches to Address the Competitiveness Concerns of Mitigating Greenhouse Gas Emissions," *National Tax Journal*, 2017.

APPLICATIONS FOR CARBON TAX REVENUE

Although a tax may be levied on fossil fuels or GHG emission sources at various points in the economy, the carbon tax impacts may be experienced elsewhere.[52] Policymakers have multiple options to address these expected impacts. Policymakers would face challenging decisions regarding the distribution of the new carbon tax revenues. As discussed below, some economic analyses indicate that certain distributions of tax revenue—depending on the level of the tax— would have a greater economic impact than the direct effects from the tax or fee on GHG emissions.[53]

Carbon tax revenues could be treated as general fund revenue without a dedication to a specific purpose in the enacting legislation (i.e., subject to the annual appropriations process), or policymakers could state that the new revenues would support deficit (or debt) reduction. Alternatively, the enacting legislation could return the tax revenue to the economy in some manner, sometimes called "revenue recycling." All of the carbon tax legislative proposals in recent Congresses have proposed some manner of revenue recycling, specifically directing the carbon tax revenue to support specific policy objectives.[54]

Carbon tax revenues may be used to support a variety of policy goals. When deciding how to allocate the new revenue stream, policymakers would likely encounter trade-offs among objectives, including:

- reducing the economy-wide costs resulting from a carbon tax program;

---

[52] In economic literature, this measure is referred to as the tax incidence.

[53] See, for example, Jared C. Carbone et al., *Deficit Reduction and Carbon Taxes: Budgetary, Economic, and Distributional Impacts*, Resources for the Future, 2013; Adele Morris and Aparna Mathur, "A Carbon Tax in Broader U.S. Fiscal Reform: Design and Distributional Issues," Center for Climate and Energy Solutions, 2014.

[54] See CRS Report R45472, *Market-Based Greenhouse Gas Emission Reduction Legislation: 108th Through 115th Congresses*, by Jonathan L. Ramseur.

- alleviating the costs borne by subgroups in the U.S. population, particularly low-income households and/or communities most dependent on carbon-intensive economic activity; and
- supporting specific policy objectives, such as domestic employment, climate change adaptation, energy efficiency, technological advance, energy diversity, or federal deficit reduction, among others.

In general, economic carbon tax studies have found that the relative ranking of revenue recycling options to mitigate the economy-wide impacts is generally the opposite of the relative ranking for alleviating distributional impacts. The contrasting relative rankings highlight a central tradeoff policymakers would face when deciding how to allocate carbon tax revenues.

The following sections discuss these trade-offs and some of the revenue application options that have received attention in recent years. A large body of economic literature has examined the economic impacts of hypothetical carbon tax programs, particularly the impacts of using the carbon tax revenues for different purposes. Many of the economic studies cited below were prepared prior to the enactment of the Tax Cuts and Jobs Act (TCJA, P.L. 115-97). Signed by President Trump in December 2017, the act changed various elements of the U.S. federal tax system.[55] In particular, the act lowered the corporate income tax rate from 35% to 21%. As discussed below, adjusting the corporate income tax rate is one of the central policy options generally considered in carbon tax economic literature both before and after enactment of P.L. 115-97. Based on a selected review of the economic literature that includes the tax code changes in P.L. 115-97, the central conclusions from carbon tax literature regarding revenue recycling appear to be largely unchanged.[56]

---

[55] See CRS Report R45145, *Overview of the Federal Tax System in 2018*, by Molly F. Sherlock and Donald J. Marples.

[56] See, for example, Lawrence Goulder et al., *Impacts of a Carbon Tax Across US Household Income Groups: What are the Equity-Efficiency Trade-Offs?*, Resources for the Future, 2018; John W. Diamond and George R. Zodrow, *The Effects of Carbon Tax Policies on the Economy and the Welfare of Households*, Columbia University, SIPA Center on Global

Economic Analyses Typically Do Not Include Policy Benefits

In general, carbon tax analyses do not consider the benefits that would be gained by reducing GHG emissions and avoiding climate change and its adverse impacts. Rather, most examine only the cost side of a cost-benefit analysis. This is largely because most economic models are structured to simulate markets and goods and services valued in markets, not effects on human health, species, and the environment.

In addition, carbon tax analyses generally do not include potential ancillary benefits that reduced GHG emissions could provide. For example, a reduction in GHG emissions from certain sectors may entail a reduction in hazardous air pollutants, which could provide health-related benefits.[57]

Typically, when this and many other reports discuss the "costs" of a policy, costs refer only the gross costs and not the net costs taking into account the benefits of the policy. If a carbon tax's benefits were to exceed its costs, the policy would have net benefits. Net costs or benefits could not be determined without analysis that fully includes benefits. In addition, the potential benefits of GHG emissions reduction may not accrue to the entities that bear the costs of the carbon tax.

Scientists and economists generally examine and monetize the costs of climate change—alternatively, the benefits of reducing GHGs—separately from specific carbon price policy proposals. For example, a 2015 report from EPA estimated the physical and monetary benefits to the United States of reducing global greenhouse gas emissions.[58]

Economy-Wide Impacts

A primary concern with a carbon tax is the potential economy-wide costs that may result.[59] Generally, a tax or fee on GHG emissions or the fuels that generate them would increase certain energy prices, namely

Energy Policy, 2018; Joseph Rosenberg, *Distributional Implications of a Carbon Tax*, SIPA Center on Global Energy Policy, 2018; Chen and Hafstead, "Using a Carbon Tax to Meet U.S. International Climate Pledges."

[57] In the context of EPA's Clean Power Plan rulemaking and more recently proposed Affordable Clean Energy substitute rule, whether to include co-benefits in calculated net benefits has generated controversy. See CRS Report R44341, *EPA's Clean Power Plan for Existing Power Plants: Frequently Asked Questions*, by James E. McCarthy et al.; and CRS Report R45119, *EPA's Proposal to Repeal the Clean Power Plan: Benefits and Costs*, by Kate C. Shouse.

[58] EPA, Climate Change in the United States: Benefits of Global Action, 2015, https://www.epa.gov/cira. See also GAO, Information on Potential Economic Effects Could Help Guide Federal Efforts to Reduce Fiscal Exposure, 2017, https://www.gao.gov/assets/690/687466.pdf.

[59] See, for example, the House-passed resolution "expressing the sense of Congress that a carbon tax would be detrimental to the United States economy" (July 19, 2018, H.Con.Res. 119); and letter from House Speaker Paul Ryan and Majority Leader Kevin McCarthy from American Energy Alliance et al., July 9, 2018, https://www.americanenergyalliance.org/wp-content/uploads/2018/07/CarbonTaxLetterUpdated.pdf.

fossil fuels, in the near- to medium-term as well as the prices of goods and services produced using these materials, like electricity. This outcome is inherent to the carbon tax, as its purpose is to increase the relative price of the more carbon-intensive energy sources compared to less carbon-intensive alternatives, encourage innovation in less carbon-intensive technologies, and promote other activity (e.g., energy efficiency) that may decrease emissions. These expected outcomes will have some economy-wide impacts.

Ultimately, the economy-wide effects would depend on a number of factors, including, but not limited to, the magnitude and scope of the carbon tax and, most importantly, use of the ensuing revenues. Economy-wide costs (referred to as macroeconomic costs) are often measured in terms of changes in projected gross domestic product (GDP) or another societal-scale metric, such as economic welfare.[60] The magnitude of macroeconomic impacts from a carbon tax has been a subject of debate among policymakers and stakeholders. In addition, results of economy-wide impacts will not include comparisons of impacts to different subpopulations or geographic regions, which may be of interest to policymakers.

Multiple economic studies and models have examined and compared various options for addressing the economy-wide impacts that may result from a carbon tax. One option for reducing the economic cost of a carbon tax is using the revenue to reduce existing taxes, such as those on labor, income, and investment. Economists generally describe such taxes as distortionary, because the taxes discourage economically beneficial activity, such as employment and investment.

Another option for policymakers is to use the tax revenues to address the national debt. Fewer studies have examined deficit reduction scenarios, because "modeling the effects of budget deficits is much more difficult

---

[60] *Economic welfare* may be defined as "a notion of household well-being encompassing everything that individuals value—market goods and services measured by GDP plus nonmarket items (e.g., children at home, leisure time)." Glossary in Parry et al., *Implementing a U.S. Carbon Tax*.

than modeling the effects of tax cuts."[61] Some studies have concluded that using tax revenues for this purpose would help alleviate economy-wide costs from a carbon tax because of the reduced need to impose distortionary taxes in the future.[62] These studies indicate that the economy-wide benefit would be delayed and its realization assumes policymakers would, sometime in the future, address the deficit by raising taxes.

Many recent legislative proposals would distribute the carbon tax revenue back to households in lump-sum payments. Policymakers have generally included this carbon tax revenue application to address distribution impacts (discussed below). These payments could take multiple forms. Economic analyses typically assume an equal payment to individuals or households regardless of their income or location or the effects of the carbon price on them individually. Alternatively, payments could be targeted or scaled to different segments of the population.

Among the options mentioned above, economic studies indicate that using carbon tax revenues to offset reductions in existing, distortionary taxes would be the most economically efficient use of the revenues and yield the greatest benefit to the economy overall. This concept is sometimes referred to as a "tax swap."

Using carbon tax or fee revenues to offset other distortionary taxes (e.g., labor or capital) may yield a "double-dividend,"[63] which includes:

- reduced GHG emissions; and
- reduced market distortions by reducing other distortionary taxes, such as investment or income.

---

[61] Robertson Williams and Casey J. Wichman, "Macroeconomic Effects of Carbon Taxes," in *Implementing a U.S. Carbon Tax: Challenges and Debates*, ed. Ian Parry et al., 2015.

[62] See, for example, Diamond and Zodrow, *The Effects of Carbon Tax Policies on the U.S. Economy and the Welfare of Households*; Jared C. Carbone et al., *Deficit Reduction and Carbon Taxes*; Ian Parry and Robertson C. Williams III, *Moving U.S. Climate Policy Forward: Are Carbon Taxes the Only Good Alternative?*, Resources for the Future, 2011; Warwick McKibben et al., *The Potential Role of a Carbon Tax in U.S. Fiscal Reform*, Brookings Institution, 2012.

[63] For further information, see Dale W. Jorgenson et al., *Double Dividend: Environmental Taxes and Fiscal Reform in the United States* (Cambridge, MA: MIT Press, 2013); and Jaume Freire-González, "Environmental Taxation and the Double Dividend Hypothesis in CGE Modelling Literature: A Critical Review," *Journal of Policy Modeling*, 2018.

The economic models that examine the economic impacts of a carbon tax differ in their frameworks and underlying assumptions and often include multiple scenarios involving different uses of carbon tax revenue. In general, the economic models find that certain revenue recycling options may reduce the economy-wide carbon tax impacts but may not eliminate them entirely.[64]

Some studies cite particular economic modeling scenarios in which a carbon tax with certain revenue recycling applications would produce a net increase in GDP or economic welfare compared to a baseline scenario.[65] These results indicate that, in certain modeling conditions, the economic improvements gained by reducing existing distortionary taxes would be greater than the costs imposed by the new carbon tax (without including the intended climate benefits of the policy). For example, results from a 2018 study demonstrated a net increase in GDP, compared to baseline conditions,[66] when carbon tax revenues were used to finance proportionate reductions in labor tax rates (payroll tax).[67]

---

[64] See Goulder et al., *Impacts of a Carbon Tax across US Household Income Groups*; Chen and Hafstead, "Using a Carbon Tax to Meet U.S. International Climate Pledges"; Williams and Wichman, "Macroeconomic Effects of Carbon Taxes"; Sugandha D. Tuladhar et al., "Environmental Policy for Fiscal Reform: Can a Carbon Tax Play a Role?," *National Tax Journal*, 2015; Sebastian Rausch and John Reilly, "Carbon Taxes, Deficits, and Energy Policy Interactions," *National Tax Journal*, 2015; NERA Economic Consulting (prepared for National Association of Manufacturers), *Economic Outcomes of a U.S. Carbon Tax*, 2013; Parry and Williams, *Moving U.S. Climate Policy Forward*.

[65] See Diamond and Zodrow, *The Effects of Carbon Tax Policies on the U.S. Economy and the Welfare of Households*; James R. McFarland et al., "Overview of the EMF 32 Study on U.S. Carbon Tax Scenarios," *Climate Change Economics*, 2018; Dale Jorgenson et al., "Carbon Taxes and Fiscal Reform in the United States," *National Tax Journal*, 2015; Carbone et al., *Deficit Reduction and Carbon Taxes*; Parry and Williams, *Moving U.S. Climate Policy Forward*; McKibben et al., *The Potential Role of a Carbon Tax in U.S. Fiscal Reform*.

[66] See Diamond and Zodrow, *The Effects of Carbon Tax Policies on the U.S. Economy and the Welfare of Households*, 2018.

[67] In addition, several economic studies prepared before the enactment of the TCJA found that when carbon tax revenues were used to reduce marginal tax rates on capital income—which includes corporate income tax and personal income tax on interest, dividends, and capital gains—the economy-wide impacts were positive relative to the baseline. See McFarland et al., "Overview of the EMF 32 Study on U.S. Carbon Tax Scenarios"; Dale Jorgenson et al., "The Welfare Consequences of Taxing Carbon," *Climate Change Economics*, 2018; Jorgenson et al., "Carbon Taxes and Fiscal Reform in the United States"; Carbone et al., *Deficit Reduction and Carbon Taxes*; McKibben et al., *The Potential Role of a Carbon Tax in U.S. Fiscal Reform*.

In general, the economic carbon tax studies usually agree on the *relative ranking* of revenue recycling options in terms of their ability to mitigate the economy-wide impacts of a carbon tax program. The studies indicate that the approaches that use carbon tax revenue to proportionately lower existing tax rates are able to mitigate more of the carbon tax economy-wide costs than using the revenue to provide a lump-sum distribution to individuals or households.

Researchers prepared multiple carbon tax analyses prior to the enactment of the TCJA in 2017 that estimated the magnitude GDP impacts.[68] As with other estimates relating to carbon tax impacts, the results depend on the scope of the carbon tax, underlying assumptions in the analytical model, and the terms of measurement: Some estimates measure GDP growth rates; others measure actual GDP.

Figure 1 illustrates the modeled GDP results from a 2018 carbon tax analysis that includes the changes made by the TCJA. This study assessed the GDP impacts under a \$50/mtCO_2e carbon tax (starting in 2020 and increasing by 2% annually) that would apply to CO_2 emissions from fossil fuel combustion and methane emissions from fossil fuel production activities.[69] The figure compares projected GDP impacts under a baseline scenario (i.e., no carbon tax) with three carbon tax revenue applications: a payroll tax rate reduction tax swap, a lump-sum distribution to households, and a scenario that would use tax revenue to reduce the national debt for 10 years and then use revenues for a lump-sum distribution to households. The figure projects GDP impacts in 2020, 2024, 2029, and 2039.

As the figure indicates, the payroll tax rate scenario would result in a 0.1% loss of GDP in the first year (2020), but would yield GDP gains in subsequent years compared to baseline. The lump-sum distribution

---

[68] See McFarland et al., "Overview of the EMF 32 Study on U.S. Carbon Tax Scenarios"; Capital Alpha Partners, *The Carbon Tax: Analysis of Six Potential Scenarios*, 2018 (although both of these analyses were published in 2018, they were prepared before the enactment of the TCJA); Jorgenson et al., "Carbon Taxes and Fiscal Reform in the United States"; Carbone et al., *Deficit Reduction and Carbon Taxes*; Parry and Williams, *Moving U.S. Climate Policy Forward*; McKibben et al., *The Potential Role of a Carbon Tax in U.S. Fiscal Reform*.

[69] The carbon tax framework is discussed in further detail in the study's companion publication: John Larsen et al., *Energy and Environmental Implications of a Carbon Tax in the United States*, SIPA Center on Global Energy Policy, 2018.

approach would yield GDP losses each year, ranging from 0.3% to 0.4% below the projected baseline. The deficit reduction approach would yield a range of GDP losses in the first 10 years—ranging from 0.4% to 0.04%—but would yield a GDP gain in 2039 (if not before), compared to baseline.

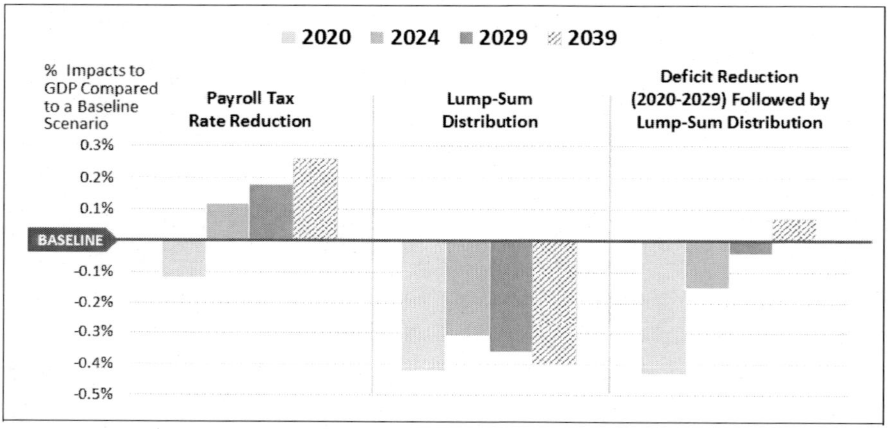

Based on a Carbon Tax of $50/mtCO2e Increasing by 2% Each Year.

Source: Data from Prepared by CRS with data from John D. Diamond and George R. Zodrow, *The Effects of Carbon Tax Policies on the U.S. Economy and the Welfare of Households*, Columbia University, SIPA Center on Global Energy Policy, 2018.

Notes: The four years included in the study and the above figure are not linear. The 2039 column has a unique color pattern to highlight the time difference between 2039 and the earlier years.

Figure 1. Estimates of GDP Impacts from a Carbon Tax with Selected Applications of Carbon Tax Revenue.

Opponents of a carbon tax approach often highlight the GDP losses that would result from a carbon tax.[70] Policymakers and stakeholders may have different perspectives regarding whether the magnitude of the GDP impacts are significant. In addition, GDP impact estimates may be presented in several ways. For example, one could compare the differences in GDP value for a particular year between carbon tax scenarios and a

[70] See, for example, National Association of Manufacturers, *Economic Outcomes of a U.S. Carbon Tax*, 2013, Executive Summary.

baseline scenario. This approach is employed in the above Figure. Alternatively, one could present the GDP losses with a cumulative measure. For instance, if one were to add up the annual GDP losses (for example, over a 10-year period) from the lump-sum scenario compared to the baseline scenario, the resulting sum would be much larger. These types of calculations would require assumptions about annual GDP growth rates.

Some may point out that the GDP impact estimates do not account for the environmental and public health benefits for reducing GHG emissions and that the GDP projections should be compared with the climate benefits achieved from the program as well as the estimated costs of taking no action. As discussed above, estimates of climate-related benefits and costs often contain considerable uncertainty and have generated debate in recent years.

Household Impacts

Many economic analyses have found that a carbon tax (before revenue recycling) would produce a regressive outcome among households, with lower-income households facing a larger impact from the tax than higher-income households.[71] However, "the degree to which a carbon tax is found to disproportionately burden low-income households varies across studies, based on the metrics against which analysts measure costs."[72]

Entities that pay the carbon tax may pass its costs back to fuel producers or forward to fuel consumers. If entities pass the costs forward, consumers would face higher prices for fuels and electricity and carbon-intensive products. When the carbon tax is passed forward to consumers, lower-income households in particular would likely face a disproportionate impact (i.e., regressive outcome), because a larger percentage of their

---

[71] See Aparna Mathur and Adele Morris, "Distributional Effects of a Carbon Tax in Broadest U.S. Fiscal Reform," *Energy Policy*, 2014.

[72] Terry Dinan, *Offsetting a Carbon Tax's Costs on Low-Income Households*, Congressional Budget Office, 2012.

income is used to pay for energy needs, such as electricity, gasoline, or home heating oil.

Many economic analyses of carbon price scenarios assume that the vast majority (if not all) of the carbon tax impact is passed forward to consumers, leading to a regressive outcome.[73] On the other hand, if entities pass the costs backward to producers, the tax impacts would fall on labor through reduced wages or owners of capital through reduced returns on investment.[74] Economic models that assume this outcome produce more progressive results (absent revenue recycling), with lower-income households experiencing smaller impacts than higher-income households.[75]

The economic analyses appear to agree that the distributional effects among households (i.e., regressive vs. progressive) of a carbon tax program would be largely dependent on how the carbon tax revenues were used.[76] A number of economic studies have used models to estimate the impacts of a carbon tax across households under several revenue distribution scenarios.[77] The results vary because the studies use different modeling frameworks, carbon tax rates and scopes, underlying assumptions, and ways to measure impacts.[78]

---

[73] Mathur and Morris state that "a number of large-scale general equilibrium models (CGE models) suggest that in the short to medium run, the burden of a carbon tax will be mostly passed forward into higher consumer prices." Mathur and Morris, "Distributional Effects of a Carbon Tax in Broadest U.S. Fiscal Reform." See Williams and Wichman, "Macroeconomic Effects of Carbon Taxes"; Terry Dinan, "Offsetting a Carbon Tax's Burden on Low-Income Households," Table 7.1, in *Implementing a U.S. Carbon Tax*.

[74] Adele Morris and Aparna Mathur, "The Distributional Burden of a Carbon Tax," in *Implementing a U.S. Carbon Tax*.

[75] See John Horowitz et al., Office of Tax Analysis, *Methodology for Analyzing a Carbon Tax*, 2017.

[76] Justin Caron et al., "Distributional Implications of a National CO2 Tax in the U.S. Across Income Classes and Regions: A Multi-Model Overview," *Climate Change Economics*, 2018.

[77] See Rosenberg, *Distributional Implications of a Carbon Tax*; Goulder et al., *Impacts of a Carbon Tax across US Household Income Groups*; Rausch and Reilly, "Carbon Taxes, Deficits, and Energy Policy Interactions"; Marron et al., *Taxing Carbon: What, Why, and How*; Robertson Williams et al., "The Initial Incidence of a Carbon Tax Across Income Groups," *National Tax Journal*, 2015; Jorgenson et al., "Carbon Taxes and Fiscal Reform in the United States"; Mathur and Morris, "Distributional Effects of a Carbon Tax in Broadest U.S. Fiscal Reform."

[78] A comparison of these differences is beyond the scope of this chapter.

For example, a 2018 study assessed the impacts to household income for different household quintiles under a carbon tax of \$50/mtCO_2e, starting in 2020.[79] This study examined four revenue distribution scenarios:

1. reduce federal deficit,
2. reduce corporate income tax rate,
3. reduce payroll tax rate, and
4. provide a per-capita rebate to households.

This chapter highlights this study, because it includes carbon tax revenue applications that have generated interest in recent years. In addition, this analysis was prepared after the 2017 tax rate changes in P.L. 115-97.

Figure 2 illustrates the modeled results, which the study measured as percentage reductions to household income. Thus, negative percentages illustrated in the figure are gains to household income. The per-capita rebate approach provides the most progressive result, yielding a net benefit for the bottom three household quintiles but a net loss for the top quintile. The fourth quintile impact is zero. By comparison, the other approaches produce varying degrees of regressive outcomes while providing a net gain for wealthier groups in two particular instances. Of the four options, the payroll tax rate reduction approach estimates the smallest variance between the income quintiles, ranging from a 0.5% loss for the lowest quintile to a 0.2% gain for the fourth quintile. The fifth quintile impact is zero.

The relative ranking among options for progressivity is generally the opposite of the relative ranking for mitigating economy-wide impacts. Other economic analyses have found similar relative rankings of revenue recycling options.[80] The contrasting rankings highlight a central tradeoff

---

[79] Rosenberg, *Distributional Implications of a Carbon Tax*.

[80] See Caron et al., "Distributional Implications of a National CO2 Tax in the U.S. Across Income Classes and Regions"; John Horowitz et al., Office of Tax Analysis, *Methodology for Analyzing a Carbon Tax*, 2017 (this study also includes a scenario splitting the revenue three ways among lump-sum, labor tax rate reduction, and corporate tax rate reduction); Jorgenson et al., "The Welfare Consequences of Taxing Carbon"; Rausch and Reilly, "Carbon Taxes, Deficits, and Energy Policy Interactions."

policymakers would face when deciding how to allocate carbon tax revenues.

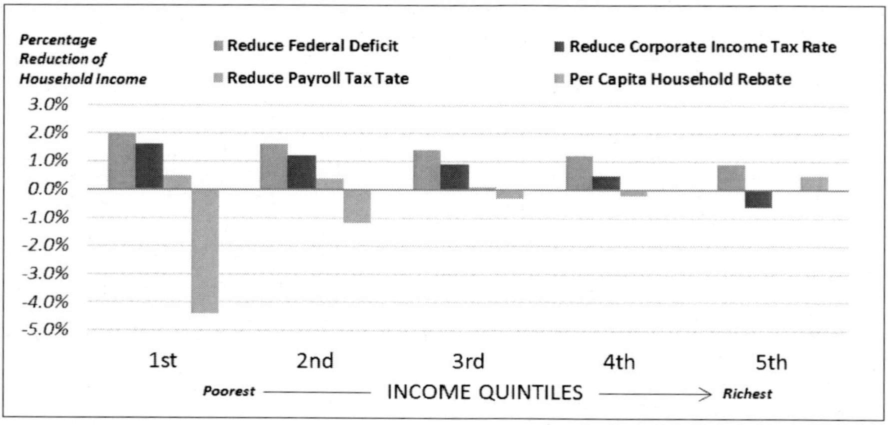

Source: Prepared by CRS using data from Joseph Rosenberg, *Distributional Implications of a Carbon Tax*, Columbia University, SIPA Center on Global Energy Policy, 2018.

Figure 2. Comparison of Carbon Tax Revenue Distribution Scenarios. Estimated Household Income Impacts in 2025.

Policymakers could allot some portion of the revenues to partially support both objectives. In a 2018 carbon tax study, economic modelers assessed a scenario in which a portion of the revenue was used to offset the welfare impacts for the lowest-income household quintile and the remaining revenue supported reductions in capital tax rates.[81] The study's models estimated that a carbon tax's impacts on the lowest-income household quintile could be counteracted with approximately 10% of the revenue. This would allow for 90% of the revenue to be used to reduce capital tax rates and thus address the economy-wide impacts from the carbon tax.

[81] Caron et al., "Distributional Implications of a National CO2 Tax in the U.S. across Income Classes and Regions."

Industry Impacts and Transition Assistance

As discussed above, a carbon tax is projected to disproportionately impact certain industries, particularly those that are described as "emission-intensive, trade-exposed industries." To address these concerns, many of the recent carbon tax legislative proposals have included design mechanisms that would attach a carbon price to certain imported materials and products (see "Border Carbon Adjustments").

Another approach to addressing the competitiveness concerns of domestic industries would involve distributing a portion of the carbon tax revenues to emission-intensive, trade-exposed industries as rebates based on their output.[82] Output rebate proposals generally determine rebate amounts by measuring emissions intensity at the relevant sector level or by a benchmark that would encourage facilities to reduce their emissions intensity.[83] These rebates could be phased out over time or continue until other nations adopt comparable carbon price policies. Under a carbon tax system in Canada, which is scheduled to take effect in 2019, industries will be subject to an "output-based pricing system."[84] Some contend that the data and administrative resources necessary to implement such a program would be substantial.[85]

A carbon tax system is also expected to disproportionately impact fossil fuel industries and the communities that rely on their employment. In particular, coal-mining communities are expected to experience substantial impacts based on the coal production declines predicted in carbon tax analyses. For example, one model estimates that under a \$50/mtCO_2e carbon tax, annual U.S. coal production would decline by almost 80% in

---

[82] As one point of reference, under H.R. 2454 (111th Congress) such industries would have received approximately 15% of the emission allowance value—analogous to 15% of carbon tax revenue—through 2025, steadily decreasing to zero thereafter.

[83] Fischer et al., "Carbon Taxes and Energy-Intensive Trade-Exposed Industries"; see also Liwayway Adkins et al., *Carbon Pricing with Output-Based Subsidies: Impacts on U.S. Industries over Multiple Time Frames*, National Center for Environmental Economics, 2012.

[84] Government of Canada, "Putting a Price on Pollution: How It Will Work," https://www.canada.ca/en/services/environment/weather/climatechange/climate-action/pricing-carbon-pollution.html.

[85] Fischer et al., "Carbon Taxes and Energy-Intensive Trade-Exposed Industries."

2030 compared to a reference case.[86] Policymakers may consider supporting worker transition or community transition assistance to help mitigate the economic impacts.[87] Several of the recent carbon tax proposals would have devoted carbon tax revenues for this objective.[88]

Other Policy Objectives

Policymakers may also consider using the carbon tax revenues to provide funding to support a range of objectives, which may include policy goals that are not directly related to climate change. Some options are identified below, and many have been included in recent legislative proposals or in state GHG mitigation programs that raise revenues:

- Technology development and deployment: Efforts to reduce the costs of emission mitigation technologies—particularly carbon capture, utilization, and sequestration—are often considered in carbon tax programs, and Congress has funded such programs in other legislation.[89]
- Energy efficiency programs: Although a carbon tax would likely stimulate energy efficiency to some degree, Congress may consider using the revenues to provide additional incentives and/or technical assistance, particularly to encourage households and small businesses to increase efficiency, which would also reduce the effects of the tax on their energy bills. States in the Regional Greenhouse Gas Initiative (RGGI) have used revenues from the program to support efficiency improvements, among other objectives.

---

[86] See Larsen et al., *Energy and Environmental Implications of a Carbon Tax in the United States*.

[87] See Adele Morris, *Build a Better Future for Coal Workers and Their Communities*, Brookings Institution, 2016.

[88] See CRS Report R45472, *Market-Based Greenhouse Gas Emission Reduction Legislation: 108th Through 115th Congresses*, by Jonathan L. Ramseur.

[89] See CRS Report R42532, *Carbon Capture and Sequestration (CCS): A Primer*, by Peter Folger.

- Biological sequestration: Trees, plants, and soils sequester carbon, removing it from the earth's atmosphere. Revenues could be used to promote carbon sequestration efforts, particularly forestry or agricultural activities, which would supplement the GHG reductions of the carbon tax.[90]
- Adaptation to climate change: Regardless of emission reduction efforts taken today, climatic changes are expected due to the ongoing accumulation of GHGs in the atmosphere. Therefore, some advocate using revenues to reduce potential damage—domestically and internationally—of a changing climate.

Allowance Value Distribution in RGGI and California's Cap-and-Trade Programs

Both the RGGI—a coalition of nine states from the Northeast and Mid-Atlantic regions—and California implement cap-and-trade programs to reduce GHG emissions. Analogous to carbon tax revenue decisions, one of the more controversial and challenging questions for policymakers when designing a cap-and-trade program is how, to whom, and for what purpose to distribute the emission allowances.

The RGGI's cap-and-trade system took effect in 2009 and applies to CO_2 emissions from electric power plants. RGGI states have answered the "how" question by employing auctions to distribute the vast majority of allowances, offering 91% of their budgeted emission allowances at auction between 2008 and 2016. As a group, RGGI states have distributed the vast majority of the emission allowance value to support energy efficiency, renewable energy, or other climate-related efforts or to provide financial assistance directly to households.[91]

California established a cap-and-trade program that took effect in 2013, covering multiple GHGs that account for approximately 85% of California's GHG emissions. In California, approximately 50% of emission allowances have been sold through an auction and 50% provided at no cost to various entities, including covered sources. Investor owned utilities (which received 16% of the allowances in 2016) are required to auction their allowances with the revenues supporting electricity consumers. California has used its cap-and-trade auction revenue to fund a variety of objectives, including a high-speed rail project, affordable housing, and low-carbon vehicles, among other programs.[92]

---

[90] In general, biological sequestration efforts are more challenging to quantify than emission reductions from other sources, such as power plants.

[91] For more details, see Table 1 in CRS Report R41836, *The Regional Greenhouse Gas Initiative: Lessons Learned and Issues for Congress*, by Jonathan L. Ramseur.

[92] For more information, see California's Legislative Analyst's Office, *The 2017-18 Budget: Cap-and-Trade*, 2017.

- Deficit reduction: The possible contribution of a carbon tax to deficit reduction would depend on the magnitude and scope of the carbon tax, various market factors, and assumptions about the size of the deficit. Some carbon tax proposals in recent congressional sessions would have allotted a portion of revenues for deficit reduction.
- Infrastructure funding: Some recent proposals have provided funding for infrastructure projects. This objective could be combined with funding for adaptation activities.

ADDITIONAL CONSIDERATIONS

Impacts on GHG Emission Levels

Multiple economic studies have estimated the emission reductions that particular carbon tax designs could achieve. Economic models provide estimates based on the best information available at the time. Comparing results from different studies is problematic, because the studies' scenarios differ in multiple ways, including the tax rate, start date, scope of the program, assumptions about economic growth and technological advances, and assumptions about other federal and state policies and their effects.

A 2018 study avoided some of these comparison difficulties by inviting modeling teams to analyze a coordinated set of scenarios.[93] The 2018 Stanford Energy Modeling Forum study ("EMF 32") assembled 11 modeling teams to analyze the economic impacts of four carbon tax scenarios starting in 2020: a $25/metric ton and $50/metric ton carbon tax, increasing annually by 1% and 5%.[94] Within each of these carbon price frameworks, the models ran separate revenue distribution scenarios: a

---

[93] For a comparison of reduction estimates from several studies, see Table 1 in Morris and Mathur, "A Carbon Tax in Broader U.S. Fiscal Reform."

[94] The EMF 32 study led to multiple papers published in a special issue in *Climate Change Economics* (March 2018). See Allen A. Fawcett et al., "Introduction to the EMF 32 Study on U.S. Carbon Tax Scenarios," *Climate Change Economics*, vol. 9, no. 1 (2018). These papers are available at https://www.worldscientific.com/toc/cce/09/01.

reduction in labor tax rates, a reduction in capital tax rates, and household rebates.

Figure 3 illustrates the study's estimates of CO_2 emissions from fossil fuel combustion.[95] The red lines in the figure display the average values for the 11 models. The shaded areas illustrate the range of results, highlighting the uncertainties in emission reduction estimates. Based on these results, the study authors concluded that each of the tax rate scenarios would likely achieve the U.S. CO_2 emission reduction targets under the Paris Agreement.[96]

As Figure 3 indicates, a carbon tax or emissions fee could be set with the expectation that it would achieve an emissions reduction target, but the resulting level of emissions would be uncertain. The uncertainty of resulting emissions may lead some stakeholders to disfavor a carbon tax or fee option to control GHG emissions. Although uncertain emissions are inherent with a carbon tax approach, Congress could employ certain design elements to enhance the emission control certainty. For example, the existing GHG emission reporting data could be used to track the impact and performance of a carbon price.[97] If policymakers determine that emission reduction is not occurring at a desired pace, the price could be amended. Legislation could establish the conditions and process by which price changes could occur.

Some may argue that adjusting the carbon price to reflect actual emissions performance would undermine the benefits of price certainty. Others may point out that unplanned adjustments to the carbon price could be politically unpalatable.

---

[95] James R. McFarland et al., "Overview of the EMF 32 Study on U.S. Carbon Tax Scenarios," *Climate Change Economics*, vol. 9, no. 1 (2018).

[96] For a description of the current U.S. commitment, see CRS In Focus IF10239, *President Obama Pledges Greenhouse Gas Reduction Targets as Contribution to 2015 Global Climate Change Deal*, by Jane A. Leggett. Although President Trump announced an intention to withdraw the United States from the agreement when eligible in 2020, no action has been taken to withdraw or modify the U.S. pledged GHG reduction contribution. See CRS In Focus IF10668, *Potential Implications of U.S. Withdrawal from the Paris Agreement on Climate Change*, by Jane A. Leggett.

[97] Starting in 2010, EPA regulations have required large emission sources and fuel suppliers (among others) to annually report GHG emissions to EPA. See EPA, "Learn About the Greenhouse Gas Reporting Program (GHGRP)," https://www.epa.gov/ghgreporting/learn-about-greenhouse-gas-reporting-program-ghgrp.

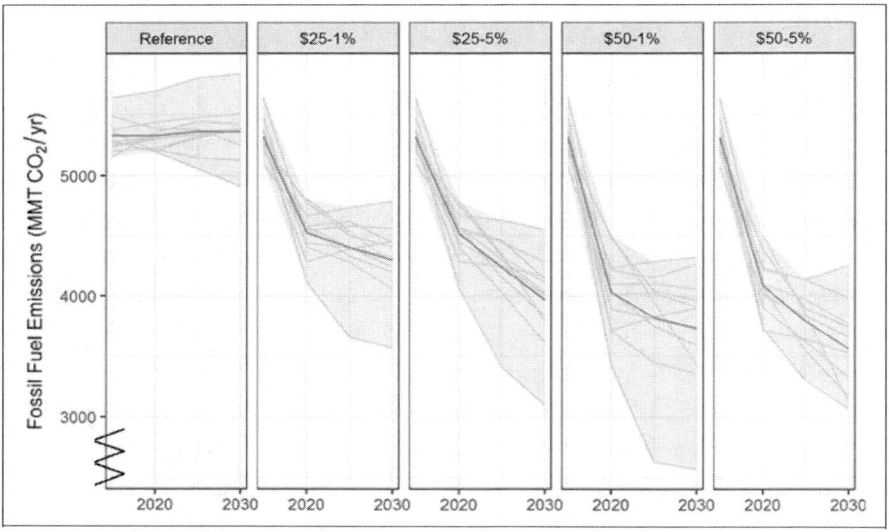

Source: Figure 1 in Alexander R. Barron et al., "Policy Insights from the EMF 32 Study on U.S. Carbon Tax Scenarios," *Climate Change Economics*, vol. 9, no. 1 (2018).

Notes: The Y-axis does not start at zero in the above figure. The model with the most aggressive reductions did not prepare results for the $50/5% scenario. Thus it appears from the figure that the $50/5% scenario has a more narrow range of emission reductions than does the $50/1% scenario.

Figure 3. Comparison of CO2 Emission Estimates from Fossil Fuel Combustion in 2020 and 2030 by Reference Case and Carbon Tax Scenarios. Results from 11 Different Modeling Groups.

For example, it may be difficult for policymakers to increase the tax rate, especially during periods of high energy prices. Some have suggested that Congress authorize an independent board or agency with the mandate to modify the tax rate administratively in order to meet pre-determined emission reduction objectives.[98] Although this approach would likely improve emission certainty, long-term price certainty may be sacrificed to

[98] See Marc Hafstead et al., "Adding Quantity Certainty to a Carbon Tax Through a Tax Adjustment Mechanism for Policy Pre-Commitment," *Harvard Environmental Law Review*, 2017; Brian Murray et al., "Increasing Emissions Certainty Under a Carbon Tax," *Harvard Environmental Law Review*, 2017; and Dieter Helm et al., "Credible Carbon Policy," in *Climate Change Policy*, ed. Dieter Helm, 2005.

some degree, depending on the authority of the delegated entity to adjust the tax rate.

Some would argue that potential year-to-year emission variations under a carbon tax would not undermine efforts to control climate change so long as long-term emission goals are achieved. Indeed, they would assert that annual emission fluctuations are preferable to price volatility that could result from an emissions cap program. They support their preference for price control by suggesting that CO_2 generates damages through its overall accumulation as concentrations in the atmosphere, not its annual flow.[99]

A potential concern of a carbon tax is whether it would be effective in reducing GHG emissions in all of its covered sectors, particularly emissions in the transportation sector. As of 2016, the transportation sector contributes the largest percentage (36%) of CO_2 emissions from fossil fuel combustion, with electric power second at 35%.[100]Carbon tax analyses generally agree that the majority of the emission reductions resulting from a carbon tax program would occur in the electricity sector. By comparison, economic models generally conclude that a carbon tax would have much less of an impact on emissions in the transportation sector.[101] Several factors explain this projected outcome. The transportation sector offers fewer opportunities to switch to less carbon-intensive fuels in the short term than does the electric power sector, which can displace coal with natural gas relatively quickly. In addition, short-term emission changes in the transportation sector are largely influenced by changes in driving demand, which has historically been relatively insensitive to gasoline price increases.[102]

---

[99] See William Pizer, *Prices vs. Quantities Revisited: The Case of Climate Change*, Resources for the Future, 1997.

[100] Historically, the electric power sector accounted for the largest percentage, but emissions in this sector have declined considerably in recent years. Between 2006 and 2016, electric power emissions decreased by 23%. EPA, *Inventory of U.S. Greenhouse Gas Emissions and Sinks: 1990–2016*, April 2018, https://www.epa.gov/ghgemissions/inventory-us-greenhouse-gas-emissions-and-sinks.

[101] See Larsen et al., *Energy and Environmental Implications of a Carbon Tax in the United States*; McFarland et al., "Overview of the EMF 32 Study on U.S. Carbon Tax Scenarios."

[102] For a further discussion see Larsen et al., *Energy and Environmental Implications of a Carbon Tax in the United States*.

Based on these projected outcomes, some may contend that to achieve deeper, long-term reductions in total GHG emissions, policymakers would need to complement a carbon tax with other programs, such as vehicle technology standards (e.g., Corporate Average Fuel Economy, CAFE) or fuel performance standards, among other options.[103]

Potential to Generate Revenues

The quantity of revenues generated under a carbon tax system depend on the program's design features, namely the tax base and rate, as well as such independent factors as prices in global energy markets. They would also depend on how covered emission sources respond to the carbon price, for example by adopting alternative technologies or changing behavior. Several carbon tax studies have prepared revenue estimates, which are presented in Table 1.

From a public finance perspective, a carbon tax may not be a reliable source of long-term funding, because a primary goal of the carbon tax is to reduce its tax base—GHG emissions. The estimates in Table 1 project carbon tax revenue values in 2020. Multiple studies have projected carbon tax revenue trajectories beyond 2020. In the 2018 EMF 32 study,[104] all but one of eight models projected carbon tax revenue increases from 2020 through 2040.[105] The carbon tax scenarios with larger annual rate increases resulted in steeper trajectories of increasing revenues through 2040. The models' estimates of annual carbon tax revenue in 2040 ranged from approximately $250 billion to $475 billion (under the tax rate scenario of $50/metric ton, increasing 5% annually).[106]

---

[103] For further information on these issues, see CRS In Focus IF10871, *Vehicle Fuel Economy and Greenhouse Gas Standards*, by Richard K. Lattanzio, Linda Tsang, and Bill Canis.

[104] Modelers in this study analyzed four carbon tax scenarios starting in 2020: a $25/metric ton and $50/metric ton carbon tax, both increasing annually by 1% and 5%.

[105] One model projected revenue declines under one of the carbon price scenarios. This model also projected a much greater utilization of CCS technology than other models. See Figure 9 in McFarland et al., "Overview of the EMF 32 Study on U.S. Carbon Tax Scenarios."

[106] See Figure 10 in McFarland et al., "Overview of the EMF 32 Study on U.S. Carbon Tax Scenarios."

Table 1. Revenue Estimates from a Carbon Tax Program Comparison of Selected Studies

Author (Year)	Scope of Program	Carbon Price	Annual Revenue Estimates
Congressional Budget Office (2018) (in nominal dollars)	Tax on CO2 emissions from energy-related activities and other selected GHG emission sources	Tax would start in 2019 at $25/mtCO2e, increasing by 2% per year plus inflation	$103 billion in 2020
EMF 32 Study (2018) (in 2010 dollars)	Tax on CO2 emissions from fossil fuels	Tax would start in 2020 at $25/mtCO2, increasing by 1% per year plus inflation	Model results ranged from $100 million to $125 million in 2020
Tax would start in 2020 at $50/mtCO2, increasing by 1% per year plus inflation	Model results ranged from $170 million to $240 million in 2020		
Office of Tax Analysis (2017) (in nominal dollars)	Tax on CO2 emissions from energy-related activities and other selected GHG emission sources	Tax would start in 2019 at $49/mtCO2e, increasing by 2% per year plus inflation	$210 billion in 2020
McKibbin et al. (2017) (in 2015 dollars)	Fee on CO2 emissions from energy-related activities	Fees start in 2020 at $27/mtCO2, increasing by 5% per year	$110 billion in 2020

Source: Congressional Budget Office, *Options for Reducing the Deficit: 2019-2028*, 2018; James R. McFarland et al., "Overview of the EMF 32 Study on U.S. Carbon Tax Scenarios," *Climate Change Economics*, 2018; John Horowitz et al., *Methodology for Analyzing a Carbon Tax*, Department of the Treasury, Office of Tax Analysis, 2017; Warwick McKibbin et al., *The Role of Border Adjustments in a U.S. Carbon Tax*, Brookings Institution, 2017.

Effects on Energy Prices and Energy Use

Fossil fuels have a wide range of CO_2 emission intensity (i.e., emissions per unit of energy).

As illustrated in Figure 4, the CO_2 emission intensity of coal is approximately 30% more than oil and approximately 80% more than natural gas. These emissions intensity differences would lead to different tax rates per unit of energy across different fuels in a carbon tax regime.

Carbon taxes could affect fuel prices in complex ways. The change in consumer fuel prices would likely not be the same as the price paid by the party directly subject to the tax. Actual price impacts for consumers would depend on multiple factors, including whether:

- a carbon tax is applied at the beginning of the production process ("upstream") to fossil fuels; and
- the price impacts are passed through to end users and not absorbed by upstream energy producers or midstream entities, such as retailers.

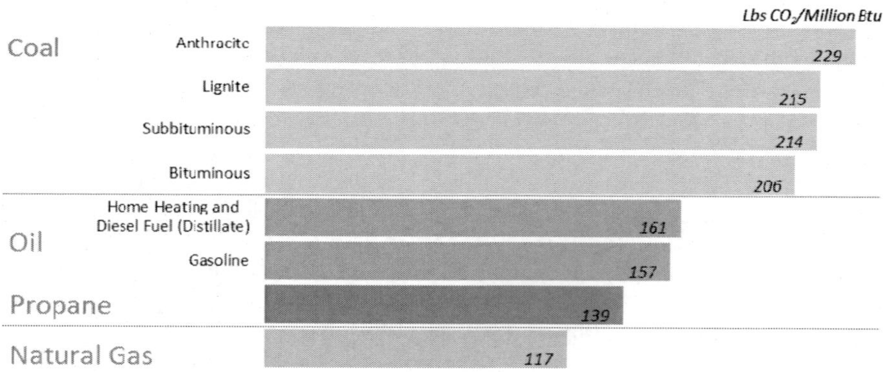

Source: Prepared by CRS; data from EIA, "Carbon Dioxide Emissions Coefficients by Fuel," https://www.eia.gov/environment/emissions/co2_vol_mass.php.

Figure 4. CO_2 Emissions Per Unit of Energy. Comparison of Selected Fossil Fuels.

In addition, market participants such as electric power plant operators can avoid paying the increased costs by substituting fuels or technologies. Energy consumers may modify their behavior in the marketplace—energy conservation, consuming less or different products and services—to mitigate impacts from the increased prices.

Table 2. Estimated Price Increases by Fuel from a Carbon Tax of $25/mt$CO_2$ on CO_2 Emissions from Fossil Fuel Combustion

Fuel	**Estimated Carbon Tax on Fuel by Volume or Weight**	**Average Market Prices (2013-2017)**	**Estimated Percentage Price Increase from Carbon Tax**
Coal	$45.00/short ton	Surface mine: $23/short ton	196%
		Underground mine: $57/short ton	79%
Natural gas	$1.25/mcf	Residential: $11/mcf	11%
		Commercial: $8/mcf	15%
		Industrial: $4/mcf	31%
		Electric power: $4/mcf	31%
Crude oil	$10.75/barrel	Domestic first purchase: $63/barrel	17%
Home heating oil	$0.20/gallon	$3.10/gallon	6%
Motor gasoline	$0.23/gallon	$2.78/gallon	8%

Source: Prepared by CRS. CRS calculated the estimated price increases for each fuel by multiplying a carbon tax rate by the CO_2 emissions intensities for each fuel. Given that carbon prices could affect fuel prices in complex ways, the actual price increases that result from the illustrative carbon taxes would depend on multiple factors. CRS generated fuel-specific emission intensities from the CO_2 coefficients (i.e., CO_2 emissions per quadrillion BTU) and thermal conversion factors (BTU per fuel unit) for each fuel. CO_2 coefficients are from EIA, "Carbon Dioxide Emission Coefficients," 2016; thermal conversion factors from EIA, Monthly Energy Review, April 2018, Appendices A2 (crude oil), A3 (home heating oil and motor gasoline), A4 (natural gas), and A5 (coal). Average market prices (2013-2017) are from EIA, Monthly Energy Review, Table 9.1 (crude oil), Table 9.10 (natural gas), and Table 9.4 (motor gasoline); home heating oil from EIA, "Weekly Heating Oil and Propane Prices"; coal from EIA, Annual Coal Report, various years, Table 28. Average coal prices include data from 2012 to 2016.

Calculated emission intensities include: Coal = 1.8 mtCO_2/short ton of coal. This value represents the CO_2 coefficient for coal (electric power sector) and the thermal conversion factor for coal consumption from the electric power sector. Surface mining accounted for 65% of total coal production in 2016. Underground mining accounted for 35%. Crude oil = 0.43 mtCO2/barrel of oil. This value represents the CO_2 coefficient for crude oil and the thermal conversion factor for "unfinished oil." Price reflects domestic first purchase price. Home heating oil = 0.008 mtCO_2/gallon of oil. Natural gas = 0.055 mtCO_2/thousand cubic feet (mcf) of natural gas. This value represents CO_2 coefficient for natural gas ("weighted national average") and the thermal conversion factor for electric power sector. Motor gasoline = 0.009 mtCO_2/gallon of gasoline. This value represents the CO_2 coefficient for "motor gasoline" and the thermal conversion factor for motor gasoline (conventional).

Table 2 includes estimates of price increases on coal, crude oil, natural gas, home heating oil, and motor gasoline based on a carbon tax rate of \$25/mtCO_2 that applies CO_2 emissions from fossil fuel combustion.

As indicated in the table, a carbon tax would have the greatest impact on the price of coal due to coal's relatively high CO_2 emissions intensity. By comparison, a carbon tax is expected to have less of an impact on the price of gasoline, increasing its price by 8%. Economic models have projected how carbon prices would impact energy use, particularly the consumption of different fossil fuels and less carbon-intensive alternatives, such as renewables or nuclear power. For example, the 2018 EMF 32 study, which included results from 11 modeling groups, assessed how several carbon tax scenarios would impact energy consumption. Highlights of these models' results (compared to reference case scenarios) include the following:[107]

- Coal consumption could decline by 40% to nearly 100% by 2030 under a \$50/mtCO_2 carbon tax, though one model projected an increase in coal due to the model incorporating CCS technology.
- Natural gas consumption estimates vary across the models, with some showing minimal change in 2030 and others showing declines ranging between 40% and 60%.
- Oil consumption estimates indicate that the largest decline (approximately 4% by 2030) would occur under the \$50/mtCO_2 carbon tax scenario.[108]
- Wind energy consumption could increase by 48% to 300% by 2030 under a \$50/mtCO_2 carbon tax scenario.

---

[107] Alexander R. Barron et al., "Policy Insights from the EMF 32 Study on U.S. Carbon Tax Scenarios," *Climate Change Economics*, vol. 9, no. 1, 2018.

[108] See Figure S4 in the supplemental materials for Barron et al., "Policy Insights from the EMF 32 Study on U.S. Carbon Tax Scenarios."

CONCLUSION

A carbon tax is one policy option to address U.S. GHG emission levels, which contribute to climate change and related impacts.

Economic modeling indicates that a carbon tax would achieve emission reductions, the level of which would depend on which GHG emissions and sources are covered and the rate of the carbon tax.

A carbon tax would generate a new revenue stream. The magnitude of the revenues would depend on the scope and rate of the tax and multiple market factors, which introduce uncertainty in the revenue projections. A 2018 CBO study estimated that a $25/metric ton tax on CO_2 emissions from energy-related activities and other selected GHG emission sources would yield approximately $100 billion in the first year of the program. To put this estimate in context, the CBO projected that total federal revenue would be $3.5 trillion in FY2019.[109]

Policymakers would face challenging decisions regarding the distribution of the new carbon tax revenues. Depending on the level of the tax, some economic analyses indicate that the distribution of tax revenue could yield greater economic impacts than the direct impacts of the tax. Some models indicate that the economic impacts are greatest in the early years of the carbon tax.

Policymakers could apply the tax revenues to support a range of policy objectives. When deciding how to allocate the revenues, policymakers would encounter trade-offs among objectives. The central trade-offs involve minimizing economy-wide costs, lessening the costs borne by specific groups—particularly low-income households—and supporting a range of specific policy objectives.

---

[109] CBO, "Options for Reducing the Deficit: 2019-2028," 2018, https://www.cbo.gov/system/files?file=2018-12/54667-budgetoptions.pdf.

A primary concern with a carbon tax is the potential economy-wide costs that may result. The potential costs would depend on a number of factors, including the magnitude, design, and use of revenues of the carbon tax. In general, economic literature finds that some of the modeled revenue applications would reduce the economy-wide costs imposed by a carbon tax but may not eliminate them entirely.

Policymakers and stakeholders may have different perspectives regarding whether these estimated economy-wide costs (typically measured in terms of GDP loss) represent a significant concern. Some argue that the estimated economy-wide costs should be compared with the policy option of not establishing a carbon tax. This comparison is uncertain as carbon tax analyses do not generally consider the benefits that would be gained by reducing GHG emissions and avoiding climate change and its adverse impacts.

Some studies cite particular economic modeling scenarios in which a carbon tax and revenue recycling could produce a net increase in GDP or economic welfare, compared to a baseline scenario. These scenarios involve using carbon tax revenues to offset reductions in existing, distortionary taxes, such as corporate income or payroll taxes. Although the models indicate that these revenue applications would yield the greatest benefit to the economy overall, the models also find that lower-income households would likely face a disproportionate impact under such revenue applications. As lower-income households spend a greater proportion of their income on energy needs, these households are expected to experience disproportionate impacts from a carbon tax if revenues were not recycled back to them in some fashion, such as a lump-sum distribution. Carbon tax revenues that are used to offset the burden imposed on various sectors or specific population groups would not be available to support other objectives.

An additional concern with a carbon tax involves potential disproportionate impacts to "emission-intensive, trade-exposed industries." Policymakers could select among several options to address these concerns, either by establishing a border carbon adjustment program or allocating some of the carbon tax revenues to selected industry sectors based on an output-based metric. If other nations were to adopt comparable carbon price policies, this concern may be alleviated to some degree.

Relatedly, a carbon tax is projected to disproportionately impact fossil fuel industries, particularly coal, and the communities that rely on their employment. To alleviate these impacts, policymakers could allocate some of the carbon tax revenue to provide transition assistance to employees or affected communities.

APPENDIX. POTENTIAL APPLICATIONS OF A CARBON TAX

Table A.1 identifies sources of GHG emissions that account for 0.5% or more of total U.S. GHG emissions. The sources are listed in descending order by their percentage contribution. CO_2 emissions from fossil fuel combustion, which accounts for almost 76% of total U.S. GHG emissions, are broken down by fossil fuel type: petroleum, coal, and natural gas.[110]

The Table identifies potential points in the economy at which a carbon tax could be applied. The Table lists the approximate number of entities that would be involved with different tax applications. The number of entities listed is current as of the most recent data available and varies accordingly by category. See Table notes for details.

The right-hand column of the Table provides additional comments for some of the emission sources. In some cases the comments discuss potential opportunities for additional GHG emissions coverage at a particular source. In other cases, the comments address potential limitations of covering all of the emissions from a particular source.

[110] An alternative breakdown of this category could include CO2 emissions by sector: electricity, transportation, industrial, residential, and commercial.

Table A.1. Selected Sources of U.S. GHG Emissions and Options for Potential Applications of a Carbon Tax

GHG Emission Source	Percentage of Total U.S. GHG Emissions (2016)[a]	Potential Tax Application Options		
		Entities	Number	Additional Comments
CO_2 from fossil fuel combustion	76			
-petroleum	34	Petroleum Option 1		With this option policymakers could levy a tax on CH4 emissions from oil field operations, potentially increasing the covered percentage of total GHG emissions from 34% to 35%.
		Petroleum wells[b] and	approx. 445,000 wells	
		Petroleum importers[c]	166	
		Petroleum Option 2		
		Petroleum refineries[d] and	137	
		Petroleum importers	166	
-coal	20	Coal Option 1		With this option policymakers could levy a tax on coal mine CH4 emissions, potentially increasing the covered percentage of total GHG emissions from 20% to 21%.
		Coal mines[e] and	690	
		Companies supplying imported coal[f]	8	
		Coal Option 2[g]		This option would only address coal used by electric utilities. Thus this option's covered percentage of total GHG emissions would be 19% instead of the 20% from "Coal Option 1."
		Coal-fired power plants[h] and	381	
		Companies supplying imported coal	8	
-natural gas	23	Natural Gas Option 1		With this option policymakers could levy a tax on CH4 emissions from "natural gas systems," potentially increasing the covered percentage of total GHG emissions from 23% to 25.5%.
		Natural gas wells[i] and	approx. 564,000 wells	
		Natural gas importers[j]	39	

GHG Emission Source	Percentage of Total U.S. GHG Emissions (2016)[a]	Potential Tax Application Options		
		Entities	Number	Additional Comments
-natural gas	23	Natural Gas Option 2[k]		This option would cover 17%-19% of total GHG emissions (compared to the 23%-25% in "Natural Gas Option 1"), because a substantial portion of natural gas that is produced domestically does not pass through a natural gas processing facility before it reaches an end user (including exported natural gas).
		Natural gas processors[l] and	551	
		Natural gas importers	39	
		Natural Gas Option 3[m]		By combining entities reporting on EIA Form 176 with natural gas fractionators, this option would cover 100% of natural gas consumption (23% of total GHG emissions). In addition, the Form 176 entities report data that could be used to levy a tax on some portion of the emissions associated with "CH4 from natural gas systems," potentially increasing the covered percentage of total GHG emissions from 23% to 24%.
		Entities reporting on EIA Form 176, including: (1) interstate/intrastate pipelines; (2) natural gas distributors; and (3) field, well, or processing plants that deliver natural gas directly to end users	1,679	
		Natural gas fractionators[n]	123	
Nitrous oxide from agricultural soils[o]	4.4	Farms[p]	approx. 2.05 million	A number of agricultural activities lead to N2O emissions, including certain soil management practices and the application of livestock manure or synthetic fertilizer. Policymakers could levy a tax on synthetic fertilizer sales to address N2O emissions from the use of synthetic fertilizers, thus covering approximately 1.0% of total U.S. GHG emissions.

Table A.1. (Continued)

GHG Emission Source	**Percentage of Total U.S. GHG Emissions (2016)[a]**	**Potential Tax Application Options**		
		Entities	Number	Additional Comments
CH4 from natural gas systems	2.5			
-field production	1.6	Natural gas wells	approx. 564,000	If policymakers levy a tax at the wellhead, the tax could potentially cover all CH4 emissions that occur at the wellhead and downstream of the wellhead: 2.5% of total GHG emissions.
-processing	0.2	Natural gas processors	551	If policymakers levy a tax at processors, the tax could potentially cover all CH4 emissions at natural gas processors and CH4 emissions that occur downstream of the processor. It is uncertain what percentage of downstream CH4 emissions would be addressed, because approximately 30% of domestically produced natural gas bypasses processing facilities (discussed in "Natural Gas Option 2").
-transmission and storage	0.5	Pipeline systems[q]	210	If policymakers levy a tax at pipeline systems, the tax could potentially cover all CH4 emissions at the pipeline level and CH4 emissions that occur downstream.
-distribution	0.2	Local distribution companies[r]	371	If policymakers levy a tax at natural gas local distribution companies (LDCs), the tax could potentially cover CH4 emissions at the LDC and any that may occur in the distribution system to end users.
Methane from livestock	2.6	Cattle operations[s]	approx. 913,000	

GHG Emission Source	Percentage of Total U.S. GHG Emissions (2016)[a]	Potential Tax Application Options		
		Entities	Number	Additional Comments
Hydrofluorocarbons from the substitution of ozone depleting substancest	2.4	Fluorinated gas manufacturers (14) and companies that import fluorinated gases (35)[u]	49	If policymakers levy a tax on fluorinated gas manufacturers and importers, the tax would not cover fluorinated gases in existing equipment. Thus the percentage of coverage would be less than 2.4%.
Methane from landfills	1.7	Municipal solid waste landfills[v]	1,900-2,000	EPA regulations have required larger landfills to address CH4 emissions since 1996.
CO_2 from nonenergy fuel use	1.7	These emissions would likely be covered by tax on fossil fuels (above)[w]		
Methane from petroleum systems[x]	0.6	Petroleum wells	approx. 360,000	
Methane from manure management	1.0	Cattle operations Swine operationsy	approx. 913,000 approx. 63,000	
Methane from coal mines	0.8	Underground coal mines (251) Surface coal mines (439)[z]	690	In 2016, approximately 76% of these CH4 emissions were from underground mines, which ventilate the CH4 for safety reasons or capture the CH4 and sell it for energy use. Attaching a tax to emissions from surface mines may present more challenges.
CO_2 from iron/steel production	0.7	Raw steel production facilities Integrated steel mills[aa]	110 11	
CO2 from cement manufacturing	0.6	Cement plants[bb]	95	

Table A.1. (Continued)

GHG Emission Source	Percentage of Total U.S. GHG Emissions (2016)[a]	Potential Tax Application Options		
		Entities	Number	Additional Comments
Percentage of Total GHG Emissions	95.0			

Source: Prepared by CRS with specific data sources provided below. Some numbers subsets may not match the total numbers due to rounding. Notes:

a. EPA, *Inventory of U.S. Greenhouse Gas Emissions and Sinks, 1990-2016*, Tables ES-2 and 3-5, April 2018.

b. Number of wells from EIA, *The Distribution of U.S. Oil and Natural Gas Wells by Production Rate, 2017 (data from 2016).* A smaller number of entities operate these wells. According to EIA, the number of well operators is approximately 15,000 (personal correspondence, May 15, 2018).

c. EIA, "Petroleum and Other Liquids, Company Level Imports" (data from February 2018). This group of companies does not include those that import "asphalt road oil," "lubricants," "wax," or "petrochemical feedstocks," which would not be subject to the above tax framework.

d. Based on the number of "operating" refineries. EIA, "Number and Capacity of Petroleum Refineries," data from 2017. Some of these refineries may produce only lubricating oils and asphalt.

e. EPA, *Inventory of U.S. Greenhouse Gas Emissions and Sinks, 1990-2016.* This figure includes underground and surface mines.

f. Eight companies supplied imported coal to power plants in December 2017. Ten power plants used imported coal in December 2017 (EIA Form EIA-923, https://www.eia.gov/electricity/data/ eia923/). In recent years, imported coal use has decreased: In 2006, the United States imported approximately 36 million tons of coal, decreasing to about 8 million tons by 2017 (EIA, *Quarterly Coal Report*, Table 4).

g. Coal Option 2 is less comprehensive than Coal Option 1, because it would not attach a carbon price to the coal that is used by entities other than electric utilities. In 2016, 95% of coal emissions were from electric utilities. The industrial sector accounted for the vast majority of the remainder. (See EPA, Inventory of U.S. Greenhouse Gas Emissions and Sinks, 1990-2016, Table 3-5.) Policymakers may consider a direct emissions approach (i.e., "downstream") for coal-fired electric utilities, because they already continuously monitor their CO2 emissions and report electronically to the EPA: Section 821 of the 1990 Clean Air Act Amendments requires electric generating facilities affected by the acid rain provisions of Title IV to monitor CO2 emissions. In addition, some argue that this approach would provide a greater stimulus to develop CCS technology. Although a price signal would be sent in either case—thus (in theory) encouraging CCS development—some contend that if coal is taxed upstream at the extraction point, some of the price signal may be weakened before it reaches coal-fired emission sources. For more information on CCS technology, see CRS Report R42532, *Carbon Capture and Sequestration (CCS): A Primer*, by Peter Folger.

h. EIA, Electric Power Annual, Table 4.1 (data from 2016).

i. Number of wells from EIA, *The Distribution of U.S. Oil and Natural Gas Wells by Production Rate*, 2017 (data from 2016).

j. EIA data indicate that 28 natural gas pipelines imported oil in 2017 ("U.S. Natural Gas Imports by Point of Entry"). In addition, there are 11 liquefied natural gas import terminals (California Energy Commission, http://www.energy.ca.gov/lng/worldwide_united_states.html).

k. CRS used 2016 data for this estimate, because 2017 data were not available for all of the data points. Although Natural Gas Option 2 would apply to fewer entities than would Natural Gas Option 1, Option 2 would cover a smaller percentage of emissions from natural gas combustion. A substantial portion of natural gas that is produced domestically does not pass through a natural gas processing facility before it reaches an end user, including exported natural gas. In 2016, this unprocessed volume was approximately 8.2 trillion cubic feet (tcf), accounting for 29% of natural gas "marketed production." CRS calculated this figure using EIA "marketed production" (28.5 tcf) and "natural gas processed" (20.3 tcf) data. To estimate the percentage of total GHG emissions covered under Option 2, CRS assumed that all of the natural gas exports (2.3 tcf in 2016) were processed, yielding 18.0 tcf processed for U.S. consumption. Combined with imported natural gas (3.0 tcf in 2016), the aggregate volume (21.0 tcf) accounts for 77% of natural gas consumption in 2016 (27.1 tcf). This equates to approximately 17% of total U.S. GHG emissions. In addition, under Option 2, policymakers could cover some portion of the CH4 emissions from "natural gas systems." This could increase the total GHG emission coverage by a range of 0.4% (processing and distribution) to 1.5% (processing, distribution, and all transmission/storage), thus yielding the 17%-19% range in the table.

l. EIA, "Natural Gas Annual Respondent Query System, Processing Capacity" (data for 2014).

m. EIA requires certain entities to submit specific natural gas data on Form EIA-176. In 2016, 1,679 entities reported natural gas deliveries (25.5 tcf in aggregate) to end users: electricity generators, industry, commercial and residential consumers. The end-use data include domestic and imported natural gas. This accounts for 94% of natural gas consumption. In addition, policymakers could attach a tax to natural gas liquids at fractionators that remove natural gas liquids from natural gas. In 2016, approximately 1.6 tcf of natural gas liquids were extracted (calculated by subtracting "dry production" from "marketed production"). Combining Form EIA-176 entities with natural gas liquid fractionators would apply a carbon price to approximately 100% of natural gas consumption, covering 23% of GHG emissions. In addition, the Form 176 entities report data that would address some portion of the emissions associated with "CH4 from natural gas systems," such as onsite use and fugitive emissions. It is uncertain what portion would be covered by the reporting entities. In this estimate CRS assumes that all of the CH4 from natural gas systems, excepting the portion attributed to production, would be addressed by applying a tax to these entities. This would add 0.9% to the coverage, thus attaching a price to approximately 24% of total GHG emissions.

n. Natural gas liquid fractionators report information to EPA pursuant to its GHG reporting program. In 2016, EPA identified 123 natural gas fractionators that reported GHG information to EPA (https://www.epa.gov/ghgreporting/ghgrp-suppliers-natural-gas-and-natural-gas-liquids).

o. Although there are no upstream transaction points that offer an opportunity to attach a price to all of these emissions, policymakers could potentially address a considerable portion by attaching a tax to the sale of synthetic fertilizers. EPA estimates that N2O emissions from the use of synthetic fertilizers account for approximately 23% of this emission source—1.0% of total U.S. GHG emissions (EPA, Inventory of U.S. Greenhouse Gas Emissions and Sinks: 1990-2016, Table 5-17, 2018). A tax on synthetic fertilizers may lead to less use, but it could "lead to other practices that release nitrogen" (Gilbert Metcalf and David

Weisbach, "The Design of a Carbon Tax," *Harvard Environmental Law Review*, vol. 33, no. 2 [2009]). In its mandatory reporting program, EPA considered requiring ammonia manufacturers and nitric acid producers to report on their synthetic fertilizer production. The agency ultimately decided to eliminate this requirement, because, according to EPA, it would not address the more than 50% of synthetic fertilizer that is imported or produced in other industries (75 Federal Register 48744, August 11, 2010).

p. USDA, *Farms and Land in Farms*, 2017 Summary, 2018.

q. Number of pipeline systems from EIA, "About U.S. Natural Gas Pipelines" (based on 2007-2008 data), https://www.eia.gov/naturalgas/archive/analysis_publications/ ngpipeline/index.html.

r. In 2016, 371 natural gas LDCs reported CO2 emissions pursuant to EPA's GHG reporting program: https://www.epa.gov/ghgreporting/ghgrp-suppliers-natural-gasand-natural-gas-liquids.

s. Cattle farm total from USDA, 2012 Census, Table 1, https://agcensus.usda.gov/Publications/.

t. The 1987 Montreal Protocol, implemented in the United States by the 1990 Clean Air Act Amendments, is phasing out ozone-depleting substances, such as chlorofluorocarbons. This development led to increased production of some substitutes, such as hydrofluorocarbons (HFCs) and perfluorocarbons, which are extremely potent GHGs. For example, HFC-134a, which accounts for the largest percentage of HFC emissions, has a global warming potential of 1,430 (see footnote 20). These materials are used in a wide variety of commercial and residential equipment, including refrigeration and air conditioning units. According to EPA, motor vehicle air conditioning units account for the largest HFC emission source. HFC emissions enter the atmosphere during "equipment manufacture and operation (as a result of component failure, leaks, and purges), as well as at servicing and disposal events" (EPA, Inventory of U.S. Greenhouse Gas Emissions and Sinks: 1990-2016). The United States and other nations agreed to phase down HFCs under the Kigali Amendment to the Montreal Protocol. The United States has not ratified the Kigali Amendment to date, but enough countries have ratified to enable the treaty to enter into force. EPA has regulated some uses of HFC under the Clean Air Act, but certain limits have been struck down in litigation. For more information, see EPA's website, https://www.epa.gov/ozone-layer-protection/recentinternational-developments-under-montreal-protocol. See also CRS In Focus IF10904, *Potential Hydrofluorocarbon Phase Down: Issues for Congress*, by Jane A. Leggett, and CRS Legal Sidebar LSB10154, *D.C. Circuit Rejects EPA's Efforts to Ban Hydrofluorocarbons: Part 1*, by Linda Tsang.

u. Under EPA's GHG reporting program, 14 facilities reported as fluorinated gas producers in 2016. This includes facilities that produce HFCs, perfluorocarbons, sulfur hexafluoride, nitrogen trifluoride, fluorinated ethers, and chlorofluorocarbons and hydrochlorofluorocarbons, including chlorodifluoromethane. (See https://www.epa.gov/ghgreporting/ghgrp-chemicals.) The number of gas importers comes from an investigation and related document prepared by the International Trade Commission (Investigation No. 731-TA-1279, Hydrofluorocarbon Blends and Components from China, 2015, p. 77, https://www.usitc.gov/publications/ 701_731/pub4558.pdf). An upstream price on fluorinated gases would not impact HFC emissions in existing equipment. Thus the percentage of coverage would be less than the total GHG contribution (2.4%). Some observers have suggested an alternative approach to pricing HFC emissions. For example, policymakers could set up a deposit-refund system at which an initial presumptive tax/fee would be applied to the manufacture or purchase of an item (e.g., vehicle air conditioning unit) and a refund would be provided upon proof of proper disposal (Gilbert Metcalf and

David Weisbach, "The Design of a Carbon Tax," *Harvard Environmental Law Review*, vol. 33, no. 2 [2009]). This approach would cover only HFC emissions from the equipment in the fee/rebate program.

v. Number of landfills from EPA, Inventory of U.S. Greenhouse Gas Emissions and Sinks: 1990-2016. For more information, see CRS Report R43860, *Methane: An Introduction to Emission Sources and Reduction Strategies*, coordinated by Richard K. Lattanzio.

w. In an upstream carbon tax system, policymakers would attach a price to fossil fuels before the fuel use. Relatively small quantities of fossil fuels are used for a wide range of nonenergy purposes, including as ingredients in plastics, asphalt, lubricants, or other products. Of the total carbon consumed for nonenergy purposes, approximately 60% is stored in the final product and not released to the atmosphere—thus the estimated 0.8% in the table. The remaining carbon is released as emissions, which may occur during the manufacturing process or during the product's use, such as solvent application (EPA, Inventory of U.S. Greenhouse Gas Emissions and Sinks: 1990-2016). Many carbon tax proposals would provide a tax credit when carbon is stored permanently in end products.

x. EPA, *Inventory of U.S. Greenhouse Gas Emissions and Sinks: 1990-2016*, Table 3-36.

y. Hog farm total from USDA, 2012 Census, Table 1, https://agcensus.usda.gov/Publications/.

z. EPA, *Inventory of U.S. Greenhouse Gas Emissions and Sinks: 1990-2016.*

aa. U.S. Geological Survey, *Mineral Commodity Summaries 2018, 2018 (data for 2017)*, https://minerals.usgs.gov/minerals/pubs/mcs/2018/mcs2018.pdf. bb. Number of facilities based on EPA's GHG reporting program data. See https://www.epa.gov/ghgreporting/ghgrp-minerals.

In: Climate Change
Editor: Iyana Hendricks
ISBN: 978-1-53617-757-2

Chapter 5

SUPERFUND: EPA SHOULD TAKE ADDITIONAL ACTIONS TO MANAGE RISKS FROM CLIMATE CHANGE*

United States Government Accountability Office

ABBREVIATIONS

CERCLA	Comprehensive Environmental Response, Compensation, and Liability Act of 1980
EPA	Environmental Protection Agency
FEMA	Federal Emergency Management Agency
National Contingency Plan	National Oil and Hazardous Substances Pollution Contingency Plan
NCA	Fourth National Climate Assessment
NOAA	National Oceanic and Atmospheric Administration

* This is an edited, reformatted and augmented version of United States Government Accountability Office; Report to Congressional Requesters, Publication No. GAO-20-73, dated October 2019.

NPL	National Priorities List
PRP	potentially responsible party
SLR	sea level rise

Why GAO Did This Study

Administered by EPA, Superfund is the principal federal program for addressing sites containing hazardous substances. EPA lists some of the most seriously contaminated sites—most of which are nonfederal—on the NPL and has recorded over 500 contaminants, including arsenic and lead, at those sites. Climate change may make some natural disasters more frequent or more intense, which may damage NPL sites and potentially release contaminants, according to the Fourth National Climate Assessment.

GAO was asked to review issues related to the impact of climate change on nonfederal NPL sites. This chapter examines, among other objectives, (1) what available federal data suggest about the number of nonfederal NPL sites that are located in areas that may be impacted by selected climate change effects and (2) the extent to which EPA has managed risks to human health and the environment from the potential impacts of climate change effects at such sites. GAO analyzed available federal data; reviewed laws, regulations, and documents; interviewed federal officials and stakeholders; visited three nonfederal NPL sites that experienced natural disasters; and compared EPA actions to manage risk to GAO's six essential elements of enterprise risk management.

What GAO Recommends

GAO is making four recommendations to EPA, including that it clarify how its actions to manage risks at nonfederal NPL sites from potential impacts of climate change align with current goals and objectives. EPA agreed with one recommendation and disagreed with the other three. GAO continues to believe that all four are warranted.

WHAT GAO FOUND

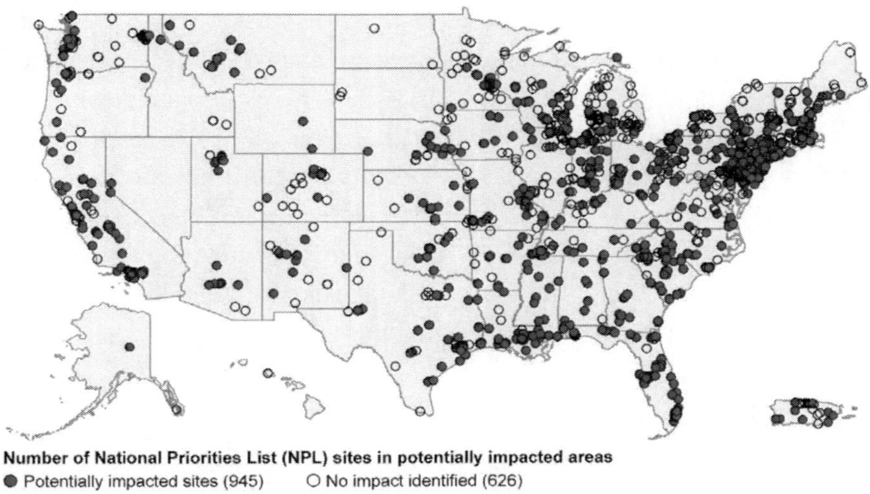

Source: GAO analysis of Environmental Protection Agency, Federal Emergency Management Agency, National Oceanic and Atmospheric Administration, and U.S. Forest Service data; MapInfo (map). | GAO-20-73

Notes: This map does not display all 1,571 active and deleted nonfederal NPL sites GAO analyzed, which also include six sites in American Samoa, the Federated States of Micronesia, Guam, the Northern Mariana Islands, and the U.S. Virgin Islands, though they are included in the counts above. Additional information on all sites GAO analyzed can be viewed at https://www.gao.gov/products/GAO-20-73. Storm surge data are not available for Alaska and Pacific islands other than Hawaii, wildfire data are not available outside the contiguous United States, and sea level rise data are not available for Alaska.

Nonfederal NPL Sites Located in Areas that May Be Impacted by Flooding, Storm Surge, Wildfires, or Sea Level Rise.

Available federal data—from the Environmental Protection Agency (EPA), Federal Emergency Management Agency, National Oceanic and Atmospheric Administration, and U.S. Forest Service—on flooding, storm surge, wildfires, and sea level rise suggest that about 60 percent of all nonfederal National Priorities List (NPL) sites are located in areas that may be impacted by these potential climate change effects. Additional information on these sites can be viewed in an interactive map and downloadable data file, available here (see figure).

EPA's actions to manage risks to human health and the environment from potential impacts of climate change effects at nonfederal NPL sites align with three of the six essential elements of enterprise risk management GAO previously identified, partially align with two essential elements, and do not align with one essential element. For example, EPA has not taken actions consistent with one essential element because it has not aligned its process for managing risks with agency-wide goals and objectives, which do not mention climate change. Without clarifying this alignment, EPA cannot ensure that senior officials will take an active role in strategic planning and accountability for managing these risks.

October 18, 2019
Congressional Requesters

In 2017, Hurricane Harvey dumped an unprecedented amount of rainfall over the greater Houston area, damaging several Superfund sites that contain hazardous substances.[1] At one site on the San Jacinto River in Texas, floodwater eroded part of the structure containing such substances, including dioxins, which are highly toxic and can cause cancer and liver and nerve damage.[2] That same year, the Fourth National Climate

---

[1] The Superfund process begins with the discovery of a potentially hazardous site or notification to the Environmental Protection Agency (EPA) of the possible release of hazardous substances, pollutants, or contaminants that may threaten human health or the environment. EPA's regional offices may discover potentially hazardous waste sites, or such sites may come to EPA's attention through reports from state agencies or citizens.

[2] According to an EPA report on the effects of the 2017 hurricanes on Superfund sites, one of several samples EPA took at the San Jacinto River Waste Pits site after Hurricane Harvey showed high dioxin concentration, above 70,000 nanograms per kilogram, which exceeds the site's risk-based cleanup level of 30 nanograms per kilogram. However, EPA officials told us that they were not able to access the site during the storm because of the severe flooding, so they could not determine whether there was a release of the dioxin- containing waste into the river at that time.

Assessment (NCA) stated that many temperature and precipitation extremes have become more frequent, more intense, or longer in duration.[3] The NCA reported that climate models are consistent with these trends continuing, which may make certain natural disasters more frequent or more intense. Further, the NCA reported that some climate change effects, including sea level rise and increased coastal flooding, could lead to the dispersal of pollutants, which could pose a risk to public health.

The Superfund program—the federal government's principal program to address sites with hazardous substances—was established by the Comprehensive Environmental Response, Compensation, and Liability Act of 1980 (CERCLA).[4] EPA is responsible for administering the program. EPA coordinates the cleanup of Superfund sites by identifying sites potentially requiring cleanup action and placing eligible sites on its National Priorities List (NPL), which includes some of the most seriously contaminated sites. As of September 2019, there were 1,336 active sites on the list, and 421 sites that EPA had determined need no further cleanup action (deleted sites). About 90 percent of these active and deleted NPL sites are nonfederal sites, where EPA generally carries out or oversees the cleanup conducted by one or more potentially responsible parties (PRP).[5] The other NPL sites—approximately 10 percent—are located at federal facilities, and the federal agencies that administer those facilities are responsible for their cleanup.[6]

---

[3] U.S. Global Change Research Program, *Climate Science Special Report, Fourth National Climate Assessment, Volume I* (Washington, D.C.: 2017). Under the Global Change Research Act of 1990, the Committee on Earth and Environmental Sciences of the Federal Coordinating Council on Science, Engineering, and Technology is to periodically prepare a scientific assessment, known as the NCA. Pub. L. No. 101-606, § 106, 104 Stat. 3096, 3101 (1990) (codified at 15 U.S.C. § 2936). The U.S. Global Change Research Program coordinates and integrates the activities of 13 participating federal departments and agencies that carry out research and support the nation's response to global change.

[4] Pub. L. No. 96-510, 94 Stat. 2767 (1980) (codified as amended at 42 U.S.C. §§ 9601-9675).

[5] Under CERCLA, PRPs generally include current or former owners or operators of a site and the generators and transporters of the hazardous substances. *See* 42 U.S.C. § 9607(a) (listing the types of parties liable for cleanup costs). In addition to EPA, other entities can be the lead agencies for cleanups under CERCLA, such as state agencies; this chapter focuses on sites for which EPA is the lead agency.

[6] Federal NPL sites are owned or operated by a department, agency, or instrumentality of the United States, such as the Departments of Defense, Energy, and the Interior. The agencies fund cleanup of federal NPL sites; this funding does not come from EPA's Superfund

CERCLA authorizes EPA to take various types of cleanup actions to prevent human and environmental exposure to contamination from nonfederal NPL sites, including remedial actions, which are long-term cleanups. As part of the Superfund cleanup process, EPA identifies, analyzes, and selects remedial actions that seek to protect human health and the environment and meet site-specific remediation goals. In September 2015, we found that annual EPA expenditures for remedial actions at nonfederal NPL sites could be considerable—about $400 million for all such sites.[7] EPA is also required to conduct or oversee reviews at least every 5 years at sites where remedial actions are complete but contaminants remain, including at deleted sites, if contaminants remaining on those sites exceed certain levels. Under EPA's regulations, the agency may take additional remedial actions to address releases at deleted sites if warranted under future conditions.[8]

Climate change may impact Superfund sites in various ways. For example, extreme precipitation events may impact Superfund sites that have contaminated sediments in aquatic environments. Specifically, in a 2007 report, the National Research Council noted that buried contaminated sediments at Superfund sites may be transported during storms or other high-flow events, becoming a source of future exposure and risk.[9] As a result of the significant risks posed by climate change and the nation's fiscal condition, in February 2013, we added *Limiting the Federal Government's Fiscal Exposure by Better Managing Climate Change Risks* to our list of areas at high risk for fraud, waste, abuse, and

---

appropriation. Processes and provisions specific to these federal sites are generally not discussed in this chapter, although they are subject to the same cleanup requirements in CERCLA.

[7] GAO, *Superfund: Trends in Federal Funding and Cleanup of EPA's Nonfederal National Priorities List Sites*, GAO-15-812 (Washington, D.C.: Sept. 25, 2015). This funding was for remedial cleanup activities, which include remedial investigations, feasibility studies, and remedial action projects (actions taken to clean up a site).

[8] 40 C.F.R. § 300.425(e)(3).

[9] National Research Council, *Sediment Dredging at Superfund Megasites: Assessing the Effectiveness* (Washington, D.C.: 2007). The National Research Council is the principal operating agency of the National Academy of Sciences and the National Academy of Engineering.

mismanagement, or most in need of transformation.[10] In March 2019, we reported on progress to address this high-risk area.[11]

You asked us to review issues related to the impact of climate change on nonfederal NPL sites. This chapter examines (1) what available federal data suggest about the number of nonfederal NPL sites that are located in areas that may be impacted by selected climate change effects; (2) the extent to which EPA has managed risks to human health and the environment from the potential impacts of climate change effects at nonfederal NPL sites; and (3) the challenges, if any, EPA faces in managing these risks.

To determine what available federal data suggest about the number of nonfederal NPL sites that are located in areas that may be impacted by selected climate change effects, we reviewed the NCA and our prior work on the Superfund program, climate change, and federal data on potential climate change effects.[12] We reviewed agency documents and interviewed officials from agencies including EPA, the Federal Emergency Management Agency (FEMA), the National Oceanic and Atmospheric Administration (NOAA), and the U.S. Forest Service. Based on our review of the NCA, we identified potential climate change effects and, based on our review of EPA documents, determined which of these effects may impact nonfederal NPL sites.[13] For these potential effects, we identified available national-level federal data sets from FEMA, NOAA, and the U.S. Forest Service on flooding, storm surge, and wildfires, respectively, which

---

[10] GAO, *High-Risk Series: An Update*, GAO-13-283 (Washington, D.C.: Feb. 14, 2013).

[11] GAO, *High-Risk Series: Substantial Efforts Needed to Achieve Greater Progress on High-Risk Areas*, GAO-19-157SP (Washington, D.C.: Mar. 6, 2019).

[12] See, for example, *Superfund Sediment Sites: EPA Considers Risk Management Principles but Could Clarify Certain Procedures*, GAO-16-777 (Washington, D.C.: Sept. 22, 2016); *Climate Change: Better Management of Exposure to Potential Future Losses Is Needed for Federal Flood and Crop Insurance*, GAO-15-28 (Washington, D.C.: Oct. 29, 2014); and *FEMA Flood Maps: Some Standards and Processes in Place to Promote Map Accuracy and Outreach, but Opportunities Exist to Address Implementation Challenges*, GAO-11-17 (Washington, D.C.: Dec. 2, 2010).

[13] These potential climate change effects, which may vary across regions, include sea level rise, which may lead to increased frequency and extent of extreme flooding from coastal storms; greater frequency and magnitude of drought; increased intensity and frequency of heavy precipitation events, which may lead to increased local flooding; increased incidence of large wildfires; and increased intensity of hurricanes.

are based on current or past conditions. The NCA reported that flooding, storm surge, and wildfires will be exacerbated by climate change in some parts of the country, but we did not identify federal data on the extent to which these effects would vary from current or past conditions. We also reviewed NOAA data on sea level rise, which provide information on the geographic extent of inundation from potential sea level rise of up to 10 feet.

In presenting the results of our analysis, we refer to flooding, storm surge, wildfires, and sea level rise as potential climate change effects. To the extent that data were available, we analyzed a range of these potential climate change effects. For example, we used the maximum extent of storm surge from Category 1 hurricanes (the lowest possible category) as well as Category 4 or 5 hurricanes (the highest possible categories), as modeled by NOAA.[14] The full range of our results can be viewed in an interactive graphic, which is available here. We focused on a range because, for three of the four effects, we had data on current hazards, which may become more intense and frequent in the future, according to the NCA. Additionally, CERCLA directs EPA to give preference to remedies that would result in the permanent and significant decrease in toxicity, mobility, or volume of the contamination. According to EPA officials, remedies at nonfederal NPL sites may have to be operational indefinitely, during which time the potential effects of climate change may become more extreme.

We obtained data from EPA's Superfund Enterprise Management System—EPA's system of record for the Superfund program—on the location and other characteristics of active and deleted nonfederal NPL sites.[15] We analyzed these data using mapping software to identify

---

[14] NOAA rates hurricanes as Categories 1 through 5 on the Saffir-Simpson Hurricane Wind Scale based on a hurricane's sustained wind speed. Hurricanes reaching Category 3 and higher are considered major hurricanes because of their potential for significant loss of life and damage. Category 4 hurricanes are those that carry sustained winds of 130–156 miles per hour. Category 5 hurricanes are those with sustained winds exceeding 156 miles per hour.

[15] According to EPA officials, EPA makes data on the location of nonfederal NPL sites available to the public strictly for informational purposes. Further, according to EPA officials, these data are not intended for use in establishing liability or calculating cost recovery statutes of limitations and cannot be relied upon to create any rights, substantive or procedural,

nonfederal NPL sites located in areas that may be impacted by selected potential climate change effects. To do so, we determined whether there are areas that may be impacted by flooding, storm surge, wildfires, and sea level rise within a 0.2-mile radius of the primary geographic coordinate of each nonfederal NPL site,[16] which we used to represent the site boundaries.[17] We reviewed this analytical approach with EPA officials. To assess the reliability of all of the data for our analysis, we, among other things, assessed the timeliness and accuracy of the data and related controls and found the data from EPA, FEMA, NOAA, and the U.S. Forest Service to be sufficiently reliable for our purposes. (See app. I for more detail on steps we took to assess the reliability of the data.)

To determine the extent to which EPA has managed risks to human health and the environment from the potential impacts of climate change effects at nonfederal NPL sites, we examined relevant provisions in CERCLA, EPA's implementing regulations, executive orders, and EPA documents on the cleanup of nonfederal NPL sites. We also reviewed relevant documents, articles, studies, and other sources that we identified by searching the websites of relevant agencies and organizations and article databases, as well as through recommendations from officials and stakeholders we interviewed. From our prior work, we identified six essential elements of risk management that we reported could help agencies anticipate and manage risk.[18]

We compared EPA's actions to manage risks to human health and the environment from the potential impacts of climate change effects with these essential elements of risk management. We analyzed information on

---

enforceable by any party in litigation with the United States or third parties. EPA officials told us that EPA reserves the right to change these data at any time without public notice.

[16] According to the *Fiscal Year 2019 Superfund Program Implementation Manual*, NPL sites in the Superfund Enterprise Management System must have one primary coordinate, which indicates the primary latitude and longitude coordinates for the site. This coordinate must be located less than 1,000 meters from the site address.

[17] In a 2018 study, EPA used a 0.2-mile radius to approximate the size of NPL sites. In this study, EPA noted that it used additional information to adjust this radius for some NPL sites. We did not make such adjustments because doing so would have required site- specific analysis, which was outside the scope of our review.

[18] GAO, *Enterprise Risk Management: Selected Agencies' Experiences Illustrate Good Practices in Managing Risk*, GAO-17-63 (Washington, D.C.: Dec. 1, 2016).

EPA actions by reviewing documents from EPA, the U.S. Global Change Research Program, and the National Research Council and our prior work on the Superfund program and climate change that we identified during our search of documents, articles, and other sources. We interviewed EPA officials at headquarters and all regional offices. We also interviewed four stakeholders and representatives of two associations with knowledge of the nexus between Superfund sites and climate changes effects to obtain their views about the extent to which EPA has managed these risks. We identified these stakeholders, such as a law professor and state environmental officials, from our search of documents, articles, and other sources and through referrals during interviews.

In addition, we selected three nonfederal NPL sites from different EPA regions to illustrate the extent to which EPA is managing these risks: the American Cyanamid site in Bridgewater, New Jersey (Region 2); the Iron Mountain Mine site near Redding, California (Region 9); and the San Jacinto River Waste Pits site in Channelview, Texas (Region 6). We selected these sites based on (1) EPA regional diversity, (2) variety in potential climate change effects, and (3) whether they had been affected by an extreme weather event within the last 10 years. The results from these illustrative examples are not generalizable to nonfederal NPL sites that we did not select. We toured each of these sites, reviewed relevant documents, and interviewed EPA officials and site stakeholders, including state and local officials, representatives of PRPs, and community organizations.

To determine the challenges EPA faces in managing risks to human health and the environment from the potential impacts of climate change effects at nonfederal NPL sites, we reviewed documents from EPA and other relevant entities, including from the National Research Council, which we obtained from EPA officials and stakeholders or identified from our search of documents, articles, and other sources. We also interviewed EPA officials at headquarters and all regional offices and stakeholders. We grouped all challenges we identified into three categories for reporting purposes: institutional, resource, and technical challenges. Appendix I describes our objectives, scope, and methodology in more detail.

We conducted this performance audit from April 2018 to October 2019 in accordance with generally accepted government auditing standards. Those standards require that we plan and perform the audit to obtain sufficient, appropriate evidence to provide a reasonable basis for our findings and conclusions based on our audit objectives. We believe that the evidence obtained provides a reasonable basis for our findings and conclusions based on our audit objectives.

BACKGROUND

CERCLA established the Superfund program to clean up contaminated sites to protect human health and the environment from the effects of hazardous substances. CERCLA requires the President to establish procedures and standards for prioritizing and responding to releases of hazardous substances, pollutants, and contaminants into the environment and to incorporate these procedures and substances into the National Oil and Hazardous Substances Pollution Contingency Plan (National Contingency Plan).[19]

Under CERCLA, PRPs are liable for conducting or paying for the cleanup of hazardous substances at contaminated sites. EPA and PRPs can undertake two types of cleanup actions: removal actions and remedial actions. Removal actions are usually short-term cleanups for sites that pose immediate threats to human health or the environment. Remedial actions are generally long-term cleanups—consisting of one or more remedial action projects—that aim to permanently and significantly reduce contamination. Remedial actions can take a considerable amount of time and money, depending on the nature of the contamination and other site-specific factors.

---

[19] According to EPA, the National Contingency Plan is the federal government's blueprint for responding to both oil spills and hazardous substance releases. The National Contingency Plan is the result of efforts to develop a national response capability and promote coordination among the hierarchy of responders and contingency plans.

EPA's Office of Superfund Remediation and Technology Innovation, which is part of the Office of Land and Emergency Management, oversees remedial actions at NPL sites, including nonfederal NPL sites.[20] At each nonfederal NPL site, the lead official who is responsible for compliance with the National Contingency Plan is the remedial project manager.[21] Management of nonfederal NPL sites is the responsibility of the EPA region in which a site is located. EPA has 10 regional offices, and each one is responsible for executing EPA programs within several states and, in some regions, territories. Figure 1 illustrates EPA's 10 regions.

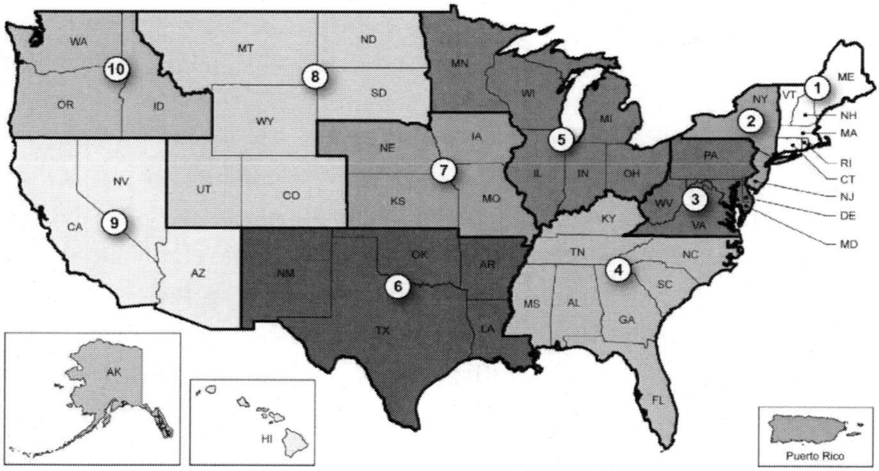

Source: GAO analysis of Environmental Protection Agency (EPA) information; MapInfo (map). | GAO-20-73

Notes: EPA Region 2 also includes the U.S. Virgin Islands. EPA Region 9 also includes American Samoa, the Federated States of Micronesia, Guam, the Northern Mariana Islands, the Republic of the Marshall Islands, and the Republic of Palau.

Figure 1. EPA's 10 Regions.

---

[20] According to EPA's website, the Office of Land and Emergency Management provides policy, guidance, and direction for the agency's emergency response and waste programs. The Office of Superfund Remediation and Technology Innovation administers the Superfund program and works to ensure that the hazardous waste sites on the NPL are cleaned up to protect human health and the environment. The Office of Land and Emergency Management was previously named the Office of Solid Waste and Emergency Response.

[21] 40 C.F.R. § 300.120(f)(1).

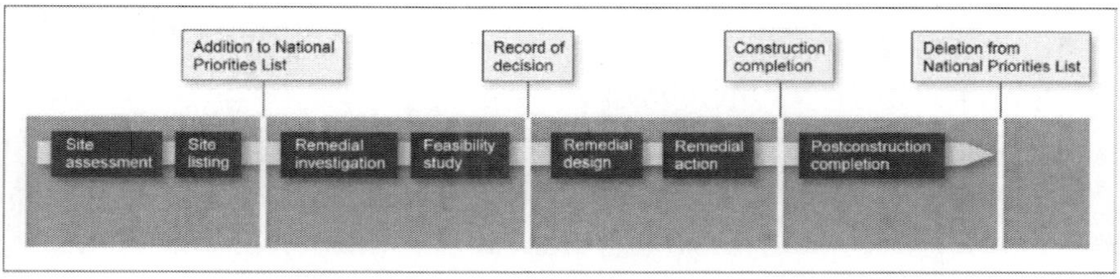

Source: GAO analysis of Environmental Protection Agency (EPA) information. | GAO-20-73

Note: EPA's actions in the remedial cleanup process may overlap, and EPA may undertake multiple actions concurrently at a site.

Figure 2. EPA's Remedial Cleanup Process at Nonfederal National Priorities List Sites.

The Superfund process begins with the discovery of a potentially hazardous site or notifications to EPA regarding the possible release of hazardous substances that may pose a threat to human health or the environment. EPA's Superfund remedial cleanup process for nonfederal NPL sites includes the actions shown in Figure 2.

- Site assessment. EPA, states, tribes, or other federal agencies evaluate site conditions to identify appropriate responses to releases of hazardous substances to the environment. During this process, EPA or other entities, such as state or tribal agencies, collect data to identify, evaluate, and rank sites using agency criteria.[22]
- Site listing. EPA considers whether to list a site on the NPL based on a variety of factors, including the availability of alternative state or federal programs that may be used to clean up the site. Sites that EPA proposes to list on the NPL are published in the *Federal Register*. After a period of public comment, EPA reviews the comments and makes final decisions on whether to list the sites on the NPL.

---

[22] EPA uses the Hazard Ranking System to guide decision-making and, as needed, to numerically assess the site's relative potential to pose a threat to human health or the environment. Sites with a Hazard Ranking System score of 28.50 or greater are eligible for listing on the NPL.

- Remedial investigation and feasibility study. EPA or the PRP will generally begin the remedial cleanup process for an NPL site by conducting a two-part study of the site. First, EPA or the PRP conducts a remedial investigation to characterize site conditions and assess the risks to human health and the environment, among other actions.[23] Second, EPA or the PRP conducts a feasibility study to assess various alternatives to address the problems identified through the remedial investigation. Under the National Contingency Plan, EPA considers nine criteria, including long-term effectiveness and permanence, in its assessment of alternative remedial actions.[24]
- Record of decision. EPA issues a record of decision that identifies its selected remedy for addressing the contamination at a site. A record of decision typically lays out the planned cleanup activities for each operable unit of the site as well as an estimate of the cost of the cleanup.[25]
- Remedial design and remedial action. EPA or the PRP plans to implement the selected remedy during the remedial design phase, and then, in the remedial action phase, EPA or the PRP carries out one or more remedial action projects.[26]
- Construction completion. EPA generally considers construction of the remedial action to be complete for a site when all physical

---

[23] As part of the remedial investigation, EPA is to identify applicable or relevant and appropriate requirements for the site. These requirements may include cleanup standards set by federal or state environmental laws that specifically address a contaminant, remedial action, location, or other circumstance at an NPL site.

[24] The nine evaluation criteria are (1) overall protection of human health and the environment; (2) compliance with applicable or relevant and appropriate requirements; (3) long-term effectiveness and permanence; (4) reduction of toxicity, mobility, or volume through treatment; (5) short-term effectiveness; (6) implementability; (7) cost; (8) state acceptance; and (9) community acceptance. 40 C.F.R. § 300.430(e)(9)(iii).

[25] An operable unit is a discrete action that comprises an incremental step toward comprehensively addressing site problems. 40 C.F.R. § 300.5. The cleanup of a site can be divided into a number of operable units, depending on the complexity of the problems associated with the site. Operable units may address geographical portions of a site, specific site problems, or initial phases of an action, or may consist of any set of actions performed over time or any actions that are concurrent but located in different parts of a site.

[26] In some cases, EPA will fund the cleanup, such as when EPA cannot identify a PRP.

construction at a site is complete, including actions to address all immediate threats and to bring all long-term threats under control.

- Postconstruction completion. EPA, the state, or the PRP performs operation and maintenance for the remedy, if needed, such as by operating a groundwater extraction and treatment system. EPA generally performs reviews of the remedy at least every 5 years to evaluate whether it continues to protect human health and the environment.[27]
- Deletion from the NPL. EPA may delete a site, or part of a site, from the NPL when the agency and the relevant state authority determine that no further site response is needed.[28]

Contaminants and Remedies at Nonfederal NPL Sites

Nonfederal NPL sites may include a variety of contaminants, and EPA may select different types of remedies to clean up the sites. EPA had recorded more than 500 contaminants at nonfederal NPL sites as of fiscal year 2014, the most recently available data. According to the Agency for Toxic Substances and Disease Registry, the highest-priority contaminants—based on a combination of their prevalence, toxicity, and potential for human exposure—are arsenic, lead, mercury, vinyl chloride, and polychlorinated biphenyls.[29] For example, in 2016, the Agency for Toxic Substances and Disease Registry reported that exposure to arsenic in drinking water is associated with various health effects, such as pulmonary

---

[27] EPA is to review Superfund remedial actions at least every 5 years where hazardous substances, pollutants, or contaminants will remain on-site above levels that allow for unlimited use and unrestricted exposure. EPA is also to report to Congress the list of sites for which these reviews are required, the results of such reviews, and any actions taken as a result of the reviews. 42 U.S.C. § 9621(c).

[28] EPA is to conduct Five-Year Reviews at deleted sites if the level of contamination at the sites exceeds what would permit unlimited use and unrestricted exposure. 40 C.F.R. § 300.430(f)(4)(ii).

[29] Under CERCLA, the Agency for Toxic Substances and Disease Registry and EPA are to prepare, in order of priority, a list of substances that are most commonly found at NPL sites and revise it periodically. The Agency for Toxic Substances and Disease Registry's most recent update to the substance priority list was in 2017. We accessed the list on April 24, 2019, at https://www.atsdr.cdc.gov/spl/.

and cardiovascular disease, diabetes, and certain cancers.[30] Contaminants may be found in different media at nonfederal NPL sites. In 2017, EPA reported that groundwater and soil were the most common contaminated media, including at the nonfederal NPL sites it analyzed.[31] To clean up a nonfederal NPL site, EPA may select various on-site or off- site remedies. For example, EPA may select on-site remedies that include treatment as well as those that do not, such as on-site containment, monitored natural recovery, and institutional controls.[32] In 2017, EPA reported that about a quarter of the decision documents for sites it analyzed included on-site treatment. EPA may also treat or dispose the contamination off-site. Examples of off-site treatment and disposal include incineration and recycling. EPA reported that sites it analyzed may have various combinations of remedies, including treatment, on-site containment, off-site disposal, and institutional controls.

Available Federal Data on Flooding, Storm Surge, Wildfires, and Sea Level Rise

Various federal agencies provide nationwide data on flooding, storm surge from hurricanes, wildfires, and sea level rise. Data on flooding, storm surge, and wildfires are generally based on current or past conditions. NOAA models the extent of inundation for various heights of sea level rise compared to the most recently available data on average high tide.

---

[30] Agency for Toxic Substances and Disease Registry, *Addendum to the Toxicological Profile for Arsenic* (Atlanta, Ga.: February 2016).

[31] Environmental Protection Agency, Office of Land and Emergency Management, *Superfund Remedy Report*, 15th Edition (July 2017). In this chapter, EPA analyzed Superfund sites, including 1,508 nonfederal NPL sites, that had signed decision documents—records of decision, amendments to records of decision, and explanations of significant differences. EPA also conducted more detailed analyses of decision documents signed in fiscal year 2012 through fiscal year 2014.

[32] Institutional controls include administrative and legal controls that minimize the potential for human exposure, for example, by limiting land use or providing information to guide behavior at the site, such as through zoning restrictions. Institutional controls are a subset of land use control, which can include physical measures such as fencing.

Flooding

FEMA provides flood hazard and risk information to communities nationwide. Among other information, FEMA provides data on coastal and riverine flooding in the National Flood Hazard Layer,[33] a database that contains the most current flood hazard data.[34] Federal law requires FEMA to assess the need to revise and update the nation's flood maps at least every 5 years.[35] Among other uses, the flood hazard data are used for flood insurance ratings and floodplain management. The National Flood Hazard Layer identifies areas at the highest risk of flooding, which are those that have a 1 percent or higher annual chance of flooding.[36] In some locations, the National Flood Hazard Layer also identifies areas with 0.2 percent or higher annual chance of flooding, which FEMA considers to be a moderate flood hazard,[37] as well as other flood hazards. The National Flood Hazard Layer also identifies areas with minimal flood hazard, including those with less than 0.2 percent annual chance of flooding, and unknown flood hazard, including areas FEMA had not assessed for flood hazards.[38] In 2018, the Technical Mapping Advisory Council noted that FEMA has produced modernized data (i.e., digital maps) for areas of the United States where 98 percent of the population resides, but has not determined the flood hazard for 40 percent of streams.[39] In general, flood hazards are

---

[33] Riverine flooding is flooding related to or caused by a river, stream, or tributary overflowing its banks because of excessive rainfall, snowmelt, or ice.

[34] FEMA provides a tool for viewing, downloading, and printing flood maps for specific locations. We accessed the tool on August 8, 2019, at https://www.fema.gov/national-flood-hazard-layer-nfhl.

[35] 42 U.S.C. § 4101(e).

[36] These areas are known as Special Flood Hazard Areas. Under federal law, in communities that participate in the National Flood Insurance Program, homeowners are required to purchase flood insurance for properties located in Special Flood Hazard Areas that are secured by mortgages from federally regulated lenders. 42 U.S.C. § 4012a(b)(1).

[37] According to the NCA, the magnitude and intensity of riverine flooding is projected to increase in the future, so areas with moderate flood hazard may have increased flood hazards in the future.

[38] We also considered areas where flood data are not available in a form compatible with our mapping software, such as those only available in paper maps, as unknown flood hazard. Paper-based maps can provide information on flood hazards to other users, such as communities and owners of Superfund sites, according to FEMA officials.

[39] The Technical Mapping Advisory Council is a federal advisory committee established to review and make recommendations to FEMA on matters related to the national flood mapping program.

based on existing conditions in the watershed and floodplains. However, in certain cases, FEMA may include flood hazard information that is based on future conditions, according to FEMA regulations.[40]

Storm Surge

NOAA provides estimates of hurricane storm surge using a model called Sea, Lake and Overland Surges from Hurricanes.[41] Estimates are available for eastern U.S. coastal areas from Texas through Maine and other areas affected by storm surge, including Hawaii, Puerto Rico, and the U.S. Virgin Islands. As of June 2019, NOAA had not modeled storm surge for the West Coast of the United States or other Pacific islands. The model takes into account a specific locale's shoreline, incorporating bay and river configurations, water depths, bridges, roads, levees, and other physical features. It estimates the maximum extent of storm surge at high tide by modeling hypothetical hurricanes under different storm conditions, such as landfall location, storm trajectory, and forward speed. NOAA models storm surge from Category 1 through Category 5 hurricanes for the Atlantic coast south of the North Carolina–Virginia border, the Gulf of Mexico, Puerto Rico, and the U.S. Virgin Islands and Category 1 through Category 4 hurricanes for the Atlantic coast north of the North Carolina–Virginia border and Hawaii.[42] According to NOAA's website, the model is to be used for educational purposes and awareness of the storm surge hazard at a city or community level.

---

[40] Future conditions refer to the flood discharges that would occur if current zoning or comprehensive land use plans were realized. *See* 44 C.F.R. § 59.1. For example, as planned buildings and parking lots are constructed, the amount of impervious land within the watershed increases, which can increase the amount of direct runoff. In our analysis, we categorized these areas as other flood hazards.

[41] According to a NOAA document, storm surge is an abnormal rise of water generated by a storm, over and above the predicted tides. Storm surge is produced by water being pushed toward the shore by the force of the winds. NOAA's storm surge hazard maps, accessed on December 31, 2018, are available at https://www.nhc.noaa.gov/nationalsurge/.

[42] According to a NOAA official, NOAA does not estimate storm surge for Category 5 hurricanes in areas where such hurricanes have not historically made landfall, such as areas north of the North Carolina–Virginia border.

In accordance with federal law, the model is also used for other purposes, such as hurricane evacuation studies.[43] According to NOAA's website, the agency updates the model for portions of the shoreline each year to account for, among other changes, new data and the addition of flood protection devices, such as levees. The model does not account for future conditions such as erosion, subsidence (i.e., the sinking of an area of land), construction, or sea level rise.

Wildfires

The U.S. Forest Service maps wildfire hazard potential based on landscape conditions and other observations.[44] According to the U.S. Forest Service, the primary intended use of the wildfire hazard potential map is to identify priority areas for hazardous fuels treatments from a broad, national- to regional-scale perspective. The U.S. Forest Service maps an index of wildfire hazard potential for the contiguous United States, based on, among other factors, annual burn probabilities and potential intensity of large fires. The U.S. Forest Service categorizes the wildfire hazard potential index into five classes of very low, low, moderate, high, and very high. For example, the U.S. Forest Service designates as "high" those areas with wildfire hazard potential index from the 85th to the 95th percentile, and "very high" above the 95th percentile. The U.S. Forest Service also categorizes areas as nonburnable (including agricultural and developed lands) and water. According to the U.S. Forest Service, areas with higher values of wildfire hazard potential represent vegetation that is more likely to burn with high intensity under certain weather conditions.

---

[43] 6 U.S.C. § 721.

[44] According to the U.S. Forest Service, the objective of the wildfire hazard potential map is to depict the relative potential for wildfire that would be difficult for suppression resources to contain. For the 2018 map, the U.S. Forest Service used spatial data sets of wildfire likelihood and intensity from 2016, spatial fuels and vegetation data from 2012, and point locations of past fire occurrence from 1992 to 2013. The U.S. Forest Service's wildfire hazard potential map, accessed on May 3, 2019, is at https://www.firelab.org/project/wildfire-hazard-potential.

However, areas with moderate, low, and very low wildfire hazard potential may still experience wildfire, particularly if they are near areas that have higher wildfire hazard potential. Wildfire hazard potential is not a forecast or wildfire outlook for any particular season as it does not include any information on current or forecasted weather or fuel moisture conditions.[45]

Sea Level Rise

NOAA models the extent of inundations from various heights of sea level rise (up to 10 feet above average high tides) for the contiguous United States, Hawaii, the Pacific islands, Puerto Rico, and the U.S. Virgin Islands and provides the results in a web mapping tool called the Sea Level Rise Viewer.[46] NOAA's guidance on the Sea Level Rise Viewer states that data are not available for Alaska. The uses of the sea level rise data include planning and education but not site-specific analysis, according to a NOAA document. NOAA labels areas as not mapped if elevation data of sufficient quality for the areas are not available. NOAA does not model natural processes, such as erosion, subsidence, or future construction, or forecast how much sea level is likely to rise in a given area. Rather, for various heights of local sea level rise, NOAA determines extent of inundation based on the elevation of an area and the potential for water to flow between areas.

Enterprise Risk Management

Enterprise risk management is a tool that allows agencies to assess threats and opportunities that could affect the achievement of their goals.

---

[45] According to a U.S. Forest Service official, the wildfire hazard potential data are not meant to substitute for local data that may more accurately capture the potential for wildfire in particular areas.

[46] NOAA's Sea Level Rise Viewer, accessed on August 7, 2019, is at https://coast.noaa.gov/digitalcoast/tools/slr.html.

In a December 2016 report, we updated our 2005 risk management framework to reflect changes to the Office of Management and Budget's Circular A-123,[47] which calls for agencies to implement enterprise risk management.[48] We also incorporated recent federal experience and identified essential elements of federal enterprise risk management.[49]

Our December 2016 report states that beyond traditional internal controls, enterprise risk management promotes risk management by considering the effect of risk across the entire organization and how it may interact with other identified risks. Additionally, it addresses other topics, such as strategy determination, governance, communicating with stakeholders, and measuring performance. The principles of enterprise risk management apply at all levels of the organization and across all functions, such as those related to managing risk to human health and the environment from the potential impacts of climate change effects at nonfederal NPL sites. The six essential elements of enterprise risk management that we identified in our December 2016 report are as follows:

1. Align risk management process with goals and objectives. Ensure that the process maximizes the achievement of agency mission and results. Agency leaders examine strategic objectives by regularly considering how risks could affect the agency's ability to achieve its mission.

---

[47] We developed our 2005 risk management framework in the context of risks associated with homeland security and combating terrorism. However, increased attention to enterprise risk management concepts and their applicability to all federal agencies and missions led us to revise our risk framework to incorporate enterprise risk management concepts. These concepts can help leaders better address uncertainties in the federal environment, including changing and more complex operating environments based on technology changes and other global factors and stakeholders seeking greater transparency and accountability.

[48] Office of Management and Budget, *Management's Responsibility for Enterprise Risk Management and Internal Control*, OMB Circular No. A-123 (July 15, 2016).

[49] GAO-17-63. See also GAO, *Risk Management: Further Refinements Needed to Assess Risks and Prioritize Protective Measures at Ports and Other Critical Infrastructure*, GAO-06-91 (Washington, D.C.: Dec. 15, 2005).

2. Identify risks. Assemble a comprehensive list of risks—both threats and opportunities—that could affect the agency's ability to achieve its goals and objectives.
3. Assess risks. Examine risks, considering both the likelihood of the risk and the impact of the risk to help prioritize risk response.
4. Respond to the risks. Select risk treatment response (based on risk appetite), including acceptance, avoidance, reduction, sharing, or transfer.
5. Monitor risks. Monitor how risks are changing and whether responses are successful.
6. Communicate and report on risks. Communicate risks with stakeholders and report on the status of addressing the risks.

About 60 Percent of Nonfederal NPL Sites are Located in Areas That may be Impacted by Selected Climate Change Effects, According to Available Data

Available federal data on flooding, storm surge, wildfires, and sea level rise suggest that about 60 percent (945 of 1,571) of all nonfederal NPL sites are located in areas that may be impacted by one or more of these potential climate change effects. These data, however, may not fully account for the number of nonfederal NPL sites that may be in such areas because (1) federal data are generally based on current or past conditions; (2) data are not available for some areas; and (3) the NCA has reported that climate change may exacerbate flooding, storm surge, and wildfires in certain regions of the United States. In addition, EPA does not have quality information on the boundaries of nonfederal NPL sites, which could affect its ability to identify the number of sites that may be impacted by one or more of these potential climate change effects.

About 60 Percent of Nonfederal NPL Sites are Located in Areas that may be Impacted by Selected Climate Change Effects; Additional Sites may be Impacted in the Future

Available federal data suggest that 945 of 1,571 nonfederal NPL sites, or about 60 percent, are located in areas that may be impacted by selected climate change effects—that is, 0.2 percent or higher annual chance of flooding or other flood hazards, storm surge from Category 4 or 5 hurricanes, high and very high wildfire hazard potential, and sea level rise of up to 3 feet. The locations of these sites are shown in Figure 3; the full results of our analysis and additional information on these sites is available in the interactive map and downloadable data file, which can be viewed at https://www.gao.gov/products/GAO-20-73.

Our analysis, however, may not fully account for the number of nonfederal NPL sites that may be impacted by the effects of climate change for various reasons. First, we represented the areas of nonfederal NPL sites based on a 0.2-mile radius around their primary geographic coordinates, which may not accurately reflect their area (i.e., they may be larger or smaller).

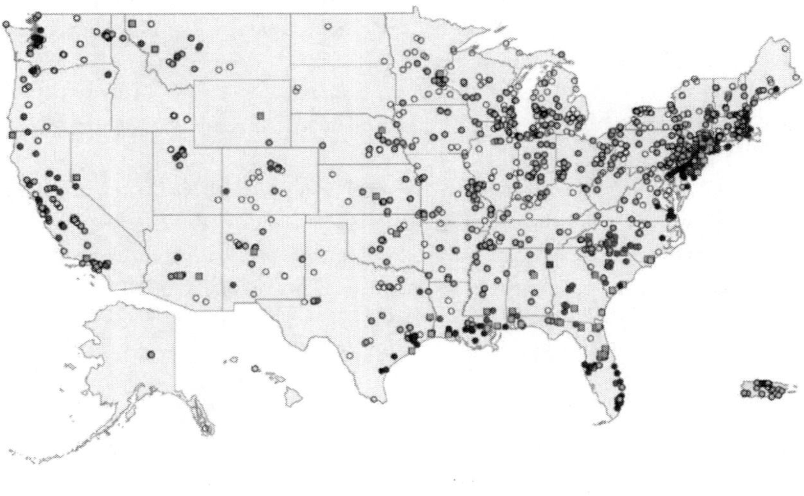

Figure 3. (Continued)

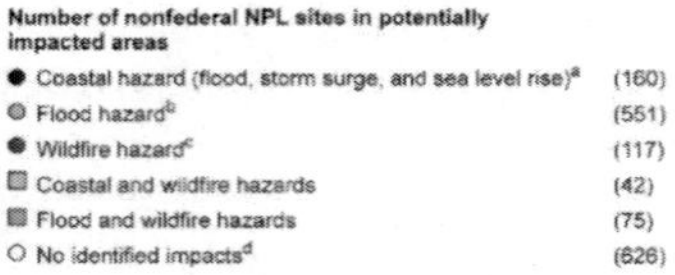

Source: GAO analysis of Environmental Protection Agency (EPA), Federal Emergency Management Agency (FEMA), National Oceanic and Atmospheric Administration (NOAA), and U.S. Forest Service data; Mapinfo (map). | GAO-20-73

Notes: We represented the boundaries of nonfederal National Priorities List (NPL) sites with a 0.2- mile radius around the primary geographic coordinate of each site. Depending on the actual site boundaries, the results of our analysis may not accurately reflect the number of nonfederal NPL sites located in these areas. This map does not display all 1,571 active and deleted nonfederal NPL sites we analyzed, which also include six sites in American Samoa, the Federated States of Micronesia, Guam, the Northern Mariana Islands, and the U.S. Virgin Islands, although we include them in the counts above. The full results of our analysis, which include additional information about these sites, are available in the interactive map and downloadable data file, which can be viewed at https://www.gao.gov/products/GAO-20-73. This analysis is based on the most recently available data from EPA, FEMA, NOAA, and the U.S. Forest Service as of March 2019.

[a] In our analysis, we use "coastal hazard" to refer to nonfederal NPL sites in areas that may be impacted by storm surge from Category 4 or 5 hurricanes and sea level rise of up to 3 feet. They may also be impacted by 0.2 percent or higher annual chance of flooding or other flood hazards. According to NOAA, storm surge data are not available for the West Coast of the United States, Alaska, and Pacific islands other than Hawaii and sea level rise data are not available for Alaska.

[b] In our analysis, we use "flood hazard" to refer to nonfederal NPL sites in areas that FEMA has determined have a 0.2 percent or higher annual chance of flooding or other flood hazards.

[c] In our analysis, we use "wildfire hazard" to refer to nonfederal NPL sites in areas with high or very high wildfire hazard potential. According to the U.S. Forest Service, wildfire hazard potential data are not available outside the contiguous United States.

[d] In our analysis, we use "no identified impacts" to refer to nonfederal NPL sites in areas with no or unknown hazard levels, including where data were unavailable. For example, FEMA flood data are not available for all locations in a format we could analyze.

Figure 3. EPA's Nonfederal NPL Sites in Areas That May Be Impacted by Flooding, Storm Surge, Wildfire, or Sea Level Rise.

We did not analyze site-specific information for these nonfederal NPL sites, including the extent of contamination and location of remedies. Such site-specific analyses would be needed to determine whether there is a risk to human health and the environment at nonfederal NPL sites as a result of these potential climate change effects.

Further, according to the NCA, EPA documents, and interviews with EPA officials, there may be other climate change effects that could impact nonfederal NPL sites, such as potential increases in salt water intrusion (the movement of saline water into freshwater aquifers), drought, precipitation, hurricane winds, and average and extreme temperatures; we did not analyze these effects because we did not identify relevant national-level federal data sets.

Flooding

We identified 783 nonfederal NPL sites—approximately 50 percent—in areas that FEMA had identified as having 0.2 percent or higher annual chance of flooding, which FEMA considers moderate flood hazard, or other flood hazards, as of October 2018.[50] Of these 783 sites, our analysis shows that 713—approximately 45 percent of all sites—are currently located in areas with 1 percent or higher annual chance of flooding, FEMA's highest flood hazard category. We provide information on the number of sites in areas with moderate or other flood hazards because, according to the NCA, heavy rainfall is increasing in intensity and frequency across the United States and is expected to continue to increase, which may lead to an increase in flooding in the future. The full results of our analysis—which include information on the sites in areas that may have 1 percent or higher annual chance of flooding, 0.2 percent or higher annual chance of flooding or other identified flood hazards, unknown flood hazard or no data, and minimal flood hazard—are available in our interactive map, which can be viewed here.

---

[50] Other flood hazards include areas with reduced risk because of levees as well as areas with flood hazard based on future conditions, for example, if land use plans were implemented. FEMA considers areas with at least 0.2 percent annual chance of flooding as having moderate flood hazard and those with 1 percent or higher annual chance of flooding to be Special Flood Hazard Areas (i.e., those with the highest chance of flooding).

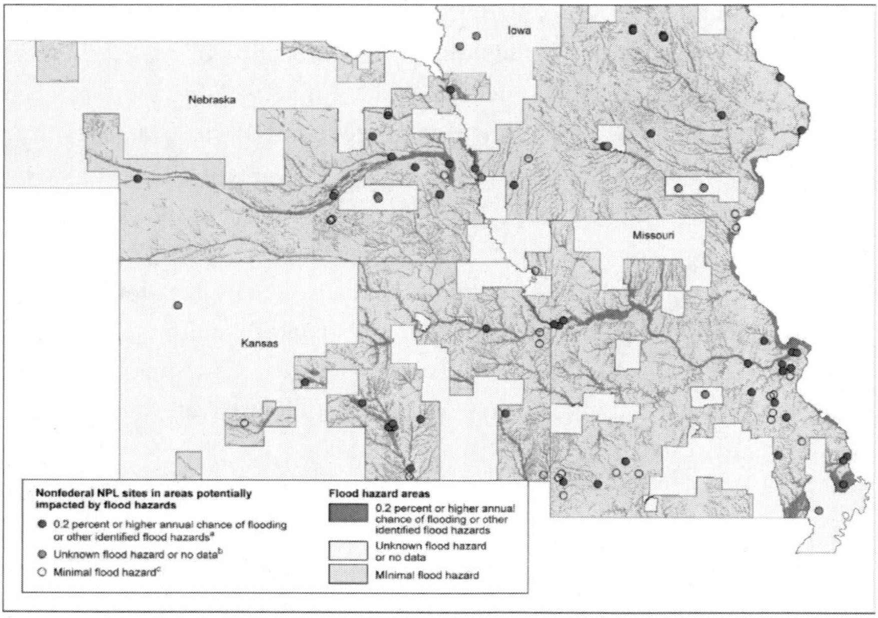

Source: GAO analysis of Environmental Protection Agency (EPA) and Federal Emergency Management Agency (FEMA) data; Mapinfo (map). | GAO-20-73

Notes: We represented the boundaries of nonfederal National Priorities List (NPL) sites with a 0.2- mile radius around the primary geographic coordinate of each site. Depending on the actual site boundaries, the results of our analysis may not accurately reflect the number of nonfederal NPL sites located in these areas. This analysis is based on EPA data as of March 2019 and FEMA data as of October 2018.

[a] FEMA considers areas with 0.2 percent or higher annual chance of flooding to have at least moderate flood hazard. Other identified flood hazards include areas with reduced risk because of levees. Of the 51 sites in these areas in EPA Region 7, 42 are located in areas that have 1 percent or higher annual chance of flooding.

[b] Unknown flood hazard or no data include areas that have flood data but that we did not analyze because the data were not available in a digital format.

[c] Minimal flood hazard includes areas with less than 0.2 percent annual probability of flooding.

Figure 4. Nonfederal NPL Sites in EPA Region 7 Located in Areas That May Be Impacted by Flooding.

For example, there are a number of nonfederal NPL sites in EPA Region 7, where states experienced record flooding in early 2019.

Specifically, as seen in Figure 4, there are 51 sites that are located in areas with 0.2 percent or higher annual chance of flooding or other identified flood hazards, of which 42 are located in areas with 1 percent or higher annual chance of flooding. Nationwide, the number of nonfederal NPL sites in areas that may be impacted by flooding currently may be higher than 783. Specifically, 217 nonfederal NPL sites are located in areas that FEMA has not assessed for flood hazards or that we did not analyze because the data were not available in a form we could use with our mapping software.

Storm Surge

We identified 187 nonfederal NPL sites—12 percent—in areas that may be inundated by storm surge corresponding to Category 4 or 5 hurricanes, the highest possible category, based on NOAA's storm surge model as of November 2018.[51] Of these sites, 102 are located in areas that may be inundated by a storm surge corresponding to Category 1 hurricanes. We analyzed areas that may be inundated by a storm surge corresponding to the highest possible category because, according to the NCA, a projected increase in the intensity of hurricanes in the North Atlantic could increase the probability of extreme flooding because of storm surge along most of the Atlantic and Gulf Coast states, beyond what would be projected based solely on relative sea level rise. However, the NCA stated that there is uncertainty in the projected increase in frequency or intensity of Atlantic hurricanes, and other factors may affect the potential for flooding because of storm surge, such as changes in overall storm frequency or tracks. The full results of our analysis, which include information on the number of sites in areas that may be inundated by storm surge from Category 1 and from Category 4 or 5 hurricanes, are available in our interactive map, which can be viewed here. In EPA Regions 2 and 3, where states experienced damage from two major hurricanes in 2017, there

---

[51] According to a NOAA website, the model does not account for future conditions, such as erosion, subsidence (i.e., the sinking of an area of land), construction, or sea level rise.

are 87 nonfederal NPL sites located within areas that may be inundated by storm surge from Category 4 or 5 hurricanes.[52]

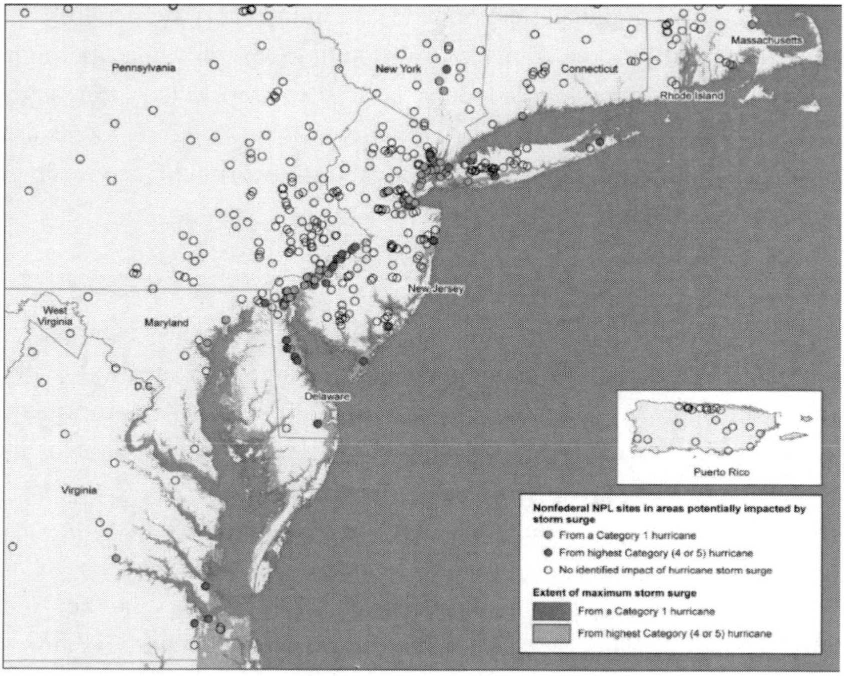

Source: GAO analysis of Environmental Protection Agency (EPA) and National Oceanic and Atmospheric Administration (NOAA) data; Mapinfo (map). | GAO-20-73

Notes: We represented the boundaries of nonfederal National Priorities List (NPL) sites with a 0.2- mile radius around the primary geographic coordinate of each site. Depending on the actual site boundaries, the results of our analysis may not accurately reflect the number of nonfederal NPL sites located in these areas. This analysis is based on EPA data as of March 2019 and NOAA data as of November 2018. EPA Region 2 also includes the U.S. Virgin Islands, which we do not depict in this figure.

Figure 5. Nonfederal NPL Sites in EPA Regions 2 and 3 Located in Areas That May Be Impacted by Storm Surge.

---

[52] Hurricanes Irma and Maria made landfall in Puerto Rico and the U.S. Virgin Islands in September 2017. These storms are considered among the five costliest hurricanes on record, according to FEMA.

Overview: The American Cyanamid site is located in Bridgewater Township, New Jersey. Prior owners used the 575-acre site for chemical and pharmaceutical manufacturing operations for more than 90 years, resulting in the contamination of soil and groundwater from 27 unlined chemical waste lagoons and containment areas, or impoundments. Contamination includes volatile and semivolatile organic compounds and metals. EPA listed the site on the National Priorities List in 1983. Wyeth Holdings LLC, a subsidiary of Pfizer Inc., acquired the site in 2009 and assumed responsibility for its cleanup.

Site status in cleanup process: Cleanup of the site is ongoing. According to EPA's website and officials, EPA is treating contaminated soil, is installing caps over three highly contaminated impoundments and all site soils, and has constructed a new water treatment system to treat contaminated groundwater. EPA is in the process of designing the remedy for impoundments 1 and 2, which contain toxic acid tar, a residual by-product of refining coal light oil.

Potential impacts of climate change: According to our analysis, the site is located in an area that has a 1 percent or higher annual chance of flooding and may be impacted by storm surge from highest possible category hurricanes. In 2011, heavy rains from Hurricane Irene flooded the site, leading to, among other impacts, a loss of power and damage to a flood control berm. Impoundments 1 and 2, located about 700 feet from the Raritan River, also flooded, as seen in the picture (below left). After the floodwaters receded, EPA inspected the berm surrounding the impoundments and conducted water sampling. EPA concluded no significant release occurred.

Actions EPA has taken to manage risks to human health and environment from impacts of climate change: Since Hurricane Irene, EPA and the responsible party (i.e., Wyeth Holdings LLC) have taken additional actions to manage risks from flooding at the site, including reinforcing flood control berms, elevating electrical equipment 5 feet higher than flood levels resulting from Hurricane Irene, and, as seen in the picture (right), installing metal pillars to protect the elevated equipment from flood debris. In the 2018 record of decision for impoundments 1 and 2, because of, among other things, risk of future flooding, EPA chose to remove and treat the acid tar off-site.

Source: GAO analysis of Environmental Protection Agency (EPA) information; EPA (left-hand photo); GAO (right-hand photo). | GAO-20-73

Figure 6, American Cyanamid National Priorities List Site—New Jersey.

Figure 5 shows these 87 sites, of which 54 sites may be inundated by storm surge from Category 1 hurricanes. Nationwide, the number of nonfederal NPL sites in areas that may be impacted by storm surge may be higher than 187 because NOAA has not modeled areas along the West Coast and Pacific islands other than Hawaii.[53] Further, our analysis did not include other potential impacts from hurricanes, such as rainfall. Figure 6 shows an example of the impact of rainfall caused by a hurricane at the American Cyanamid NPL site.

Wildfires

We identified 234 nonfederal NPL sites—15 percent—located in areas that have high or very high wildfire hazard potential—those more likely to burn with a higher intensity, based on a U.S. Forest Service model as of July 2018.

---

[53] Our analysis may not accurately account for the impacts of storm surge because we included sites in areas that are protected by levees. NOAA officials told us that storm surge in these areas is difficult to model.

For this analysis, we combined the high and very high wildfire hazard potential categories; we did not identify the number of sites in each of these categories separately. We did not analyze areas that currently have moderate or lower wildfire hazard potential because those with moderate or lower wildfire hazard potential are less likely to experience high-intensity wildfire and the extent to which wildfire hazard potential may change in the future is unknown. The full results of our analysis on the number of sites in areas with high or very high wildfire hazard potential are available in our interactive map, which can be viewed here. As seen in Figure 7, there are 22 nonfederal NPL sites in areas with high or very high wildfire hazard potential in EPA Region 9, a region that experienced wildfires in 2018, including the highly destructive Carr Fire.[54]

Nationwide, the number of nonfederal NPL sites in areas that currently have high wildfire hazard potential may be higher than 234 because wildfire hazard data are only available for the contiguous United States (i.e., there are no data for Alaska, Hawaii and other Pacific islands, Puerto Rico, and the U.S. Virgin Islands). According to the NCA, the incidence of large forest fires in the western United States and Alaska has increased since the early 1980s and is projected to further increase in those regions as the climate changes. However, the NCA noted that analyses regarding the effect of climate change on the incidence of wildfire in other parts of the United States are not readily available, so it is unknown how climate change will affect the number of nonfederal NPL sites in areas rated with high or very high wildfire hazard potential nationwide. As Figure 8 shows, wildfires can pose risks at nonfederal NPL sites, such as the Iron Mountain Mine site near Redding, California.

---

[54] The Carr Fire began on July 23, 2018, within the Whiskeytown National Recreation Area in Northern California, and by the time it was contained on August 30, 2018, it had covered approximately 229,651 acres and destroyed over 1,000 residences.

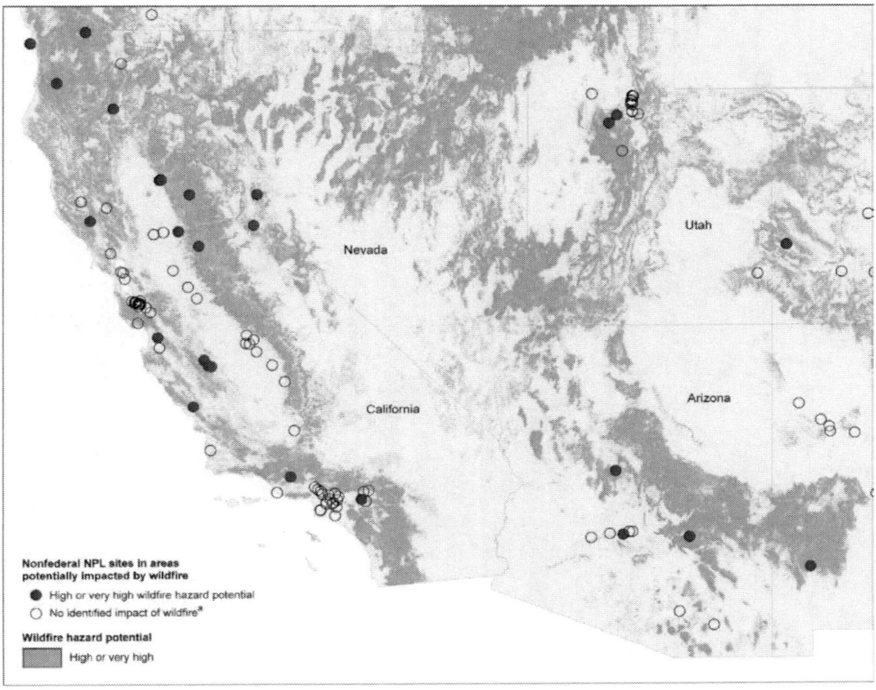

Source: GAO analysis of Environmental Protection Agency (EPA) and U.S. Forest Service data; Mapinfo (map). | GAO-20-73

Notes: We represented the boundaries of nonfederal National Priorities List (NPL) sites with a 0.2- mile radius around the primary geographic coordinate of each site. Depending on the actual site boundaries, the results of our analysis may not accurately reflect the number of nonfederal NPL sites located in these areas. This map does not depict nonfederal NPL sites that are located in areas with high or very high wildfire hazard potential in Hawaii and other Pacific islands (which are part of EPA Region 9) because the U.S. Forest Service does not model wildfire hazard potential outside the contiguous United States. This analysis is based on EPA data as of March 2019 and U.S. Forest Service data as of July 2018.

[a] No identified impact of wildfire includes unburnable areas as well as areas with very low to moderate wildfire hazard potential.

Figure 7. Nonfederal NPL Sites in EPA Region 9 Located in Areas with High or Very High Wildfire Hazard.

Overview: The 4,400-acre Iron Mountain Mine site near Redding, California, produced iron, silver, gold, copper, zinc, and pyrite through 1963. The underground mine workings and the fractured bedrock allow water and oxygen to react with the ore. The resulting acid mine drainage contains metals such as copper, cadmium, and zinc that are toxic to aquatic life, such as trout and salmon. EPA listed the site on the National Priorities List in 1983. In 2000, federal agencies and California reached a settlement with Aventis, the principal responsible party at the Iron Mountain Mine site. Global Loss Prevention, a wholly owned subsidiary of American International Group, operates the site.

Site status in cleanup process: Cleanup of the site is ongoing. EPA has constructed interim remedies, such as diverting streams to avoid contamination with acid mine drainage, and has begun a remedial investigation and feasibility study. According to EPA's sixth Five-Year Review report, in 2000, the potentially responsible party completed the construction of a water treatment system, seen in the picture, that captures most of the acid mine drainage, neutralizes it, and removes metals prior to discharge. The interim remedies remove 95 percent of the historic quantities of copper, cadmium, and zinc discharged from the Iron Mountain Mine and prevent uncontrolled releases of acid mine drainage into nearby streams and the Sacramento River in all but the most severe storms.

Potential impacts of climate change: According to our analysis, the site is located in an area with high or very high wildfire hazard potential. In July 2018, the Carr Fire burned through the site and almost destroyed the water treatment system. In the days that followed, fire was discovered in the high density polyethylene pipe that conveys acid mine drainage from one of the mines to the water treatment system. Firefighters, using specialized equipment, successfully extinguished the fire before it reached the ore body in the mine, which could have led to an explosion and substantial environmental and health hazards, according to an EPA report. EPA and state officials told us that increasing frequency and intensity of wildfires and landslides and erosion because of storm runoffs are an ongoing concern at the site.

Actions EPA has taken to manage risks to human health and environment from impacts of climate change: Following the fire, the site operator replaced portions of the pipes conveying acid mine drainage with nonflammable stainless steel, as can be seen in the bottom left corner of the picture. EPA officials told us that they plan to develop a model of water quality, including potential changing precipitation patterns because of climate change, in their remedial investigation for one of the operable units at the site.

Source: GAO analysis of Environmental Protection Agency (EPA) information; GAO (photo). | GAO-20-73

Figure 8. Iron Mountain Mine National Priorities List Site—California.

Sea Level Rise

We identified 110 nonfederal NPL sites—7 percent—located in areas that would be inundated by a sea level rise of 3 feet, based on our analysis of EPA and NOAA data as of March 2019 and September 2018, respectively. Our analysis shows that if sea level in these areas rose by 1 foot, 97 sites would be inundated. If sea level in these areas rose by 8 feet, 158 sites would be inundated. We also identified 84 nonfederal NPL sites that are located in areas that may already be inundated at high tide.[55] We provide the number of sites in areas that may be impacted by these sea level rise heights because, according to the NCA, global average sea levels are very likely to continue to rise by at least several inches in the next 15 years and by 1.0 to 4.3 feet by 2100. Further, the NCA states that a rise of as much as 8 feet by 2100 cannot be ruled out.

---

[55] These sites are located in areas at 0-foot sea level rise, which according to NOAA data is equivalent to the water level at the average of the highest of the two daily tides from 1983 to 2001.

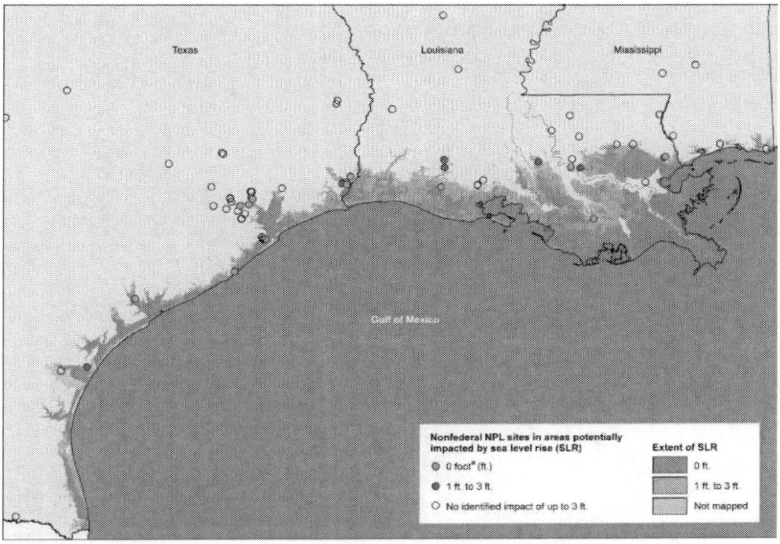

Source: GAO analysis of Environmental Protection Agency (EPA) and National Oceanic and Atmospheric Administration (NOAA) data; Mapinfo (map). | GAO-20-73

Notes: We represented the boundaries of nonfederal National Priorities List (NPL) sites with a 0.2- mile radius around the primary geographic coordinate of each site. Depending on the actual site boundaries, the results of our analysis may not accurately reflect the number of nonfederal NPL sites located in these areas. SLR is measured relative to the average highest tide from 1983 to 2001. This analysis is based on EPA data as of March 2019 and NOAA data as of September 2018.

[a] A 0 ft. SLR means an area may already be inundated at high tide.

Figure 9. Nonfederal NPL Sites in EPA Region 6 Located in Areas That Would Be Inundated by Sea Level Rise.

The full results of our analysis, which include information on the number of sites in areas that may already be inundated at high tide and that would be inundated if sea level rose by 1 foot, 3 feet, and 8 feet, are available in our interactive map, which can be viewed here. There are 23 nonfederal NPL sites located within areas that may be impacted if sea level rose by up to 3 feet in EPA Region 6, a region that has experienced land loss because of sea level rise and coastal flooding, according to the NCA.[56] In addition, as seen in Figure 9, 16—or 70 percent—of these sites may already be inundated at high tide.

Overview: The approximately 40-acre San Jacinto River Waste Pits site is located east of Houston, Texas, between two unincorporated areas known as Channelview and Highlands. In the mid-1960s, liquid and solid pulp and paper mill wastes were disposed of at the site in impoundments, or waste disposal areas. The primary hazardous substances at the site, by-products of the pulp bleaching process, are dioxins and furans, exposure to which can cause several health effects, including skin diseases and liver damage. Added to the National Priorities List in 2008, the site consists of impoundments in and adjacent to the San Jacinto River north and south of Interstate 10. As seen in the picture, the San Jacinto River covers part of the northern impoundment, the boundaries of which are marked with buoys. The International Paper Company and McGinnes Industrial Maintenance Corporation are responsible for the cleanup.

Site status in cleanup process: Cleanup of the site is ongoing. In 2010, EPA required the potentially responsible parties to, among other things, install and maintain a temporary armored cap over the waste that could withstand storm events with 1 percent or higher annual chance of occurring. The temporary armored cap includes an impervious geomembrane under the northern impoundment and a cover over the impoundment. The potentially responsible parties also stabilized and solidified part of the paper mill waste. EPA is currently designing the long-term remedy for the site.

Potential impacts of climate change: According to our analysis, the site is located in an area that has a 1 percent or higher annual chance of flooding and that may be impacted by storm surge from Category 1 hurricanes and sea level rise of 0 foot. According to the 2017 record of decision, since the installation of the temporary cap, EPA has observed repeated damage to sections of the cap, including in September 2017 from Hurricane Harvey. Record-breaking rainfall during the hurricane led to flooding, which eroded the cap in some places, exposing some of the contaminated material. EPA detected high levels of dioxins in one area it sampled.

Actions EPA has taken to manage risks to human health and environment from impacts of climate change: According to the operations, monitoring, and maintenance plan of the time-critical removal action for the site, EPA has directed the potentially responsible parties to periodically inspect the cap and conduct repairs as needed after certain flood events. In the 2017 record of decision, EPA required the potentially responsible parties to remove and treat most of the contaminated material off-site, because of, among other things, risk of future flooding from hurricanes and sea level rise.

Source: GAO analysis of Environmental Protection Agency (EPA) information; GAO (photo). | GAO-20-73

Figure 10. San Jacinto River Waste Pits National Priorities List Site—Texas.

Nationally, the number of nonfederal NPL sites that may be inundated by various heights of sea level rise will vary from the results of our analysis because different parts of the United States may experience higher or lower sea level rise than the global average. For example, the NCA states that sea level rise will be higher than the global average on the East and Gulf Coasts of the United States and lower than the global average in most of the Pacific Northwest and in Alaska. As can be seen in Figure 10, sea level rise and other coastal hazards may impact nonfederal NPL sites, such as the one in the San Jacinto River Waste Pits site in Texas, parts of which are already under water.

---

[56] There are 18 nonfederal NPL sites in EPA Region 6 that would be inundated if sea level rose by 1 foot. In addition, 28 nonfederal NPL sites would be inundated if sea level rose by 8 feet in that region.

EPA Does Not Have Quality Information on the Boundaries of Nonfederal NPL Sites

EPA does not have quality information on the boundaries of nonfederal NPL sites, which could affect its ability to identify the number of sites that may be impacted by one or more of these potential climate change effects.[57] According to EPA officials, EPA has not validated data on site boundaries and EPA's regional offices do not use a consistent geographic standard,[58] which makes it difficult to produce a national data set. In general, EPA officials told us that information on the boundaries of NPL sites has not been a focus at a national level and is not yet subject to quality standards. For example, EPA officials told us that boundary information for each NPL site represents the remedial project manager's professional judgment and remedial project managers may determine and record the boundaries of sites differently.

EPA has taken some initial actions to improve the quality of information on the boundaries of nonfederal NPL sites. In November 2017, the Office of Superfund Remediation and Technology Innovation issued a directive to all regional Superfund division directors recommending national standards for collecting and maintaining geographic information, including site boundaries.[59] EPA's 2017 directive notes that using national standards to collect geographic information, including site boundaries, promotes EPA's reporting and analytical efforts to support program implementation and evaluation. In addition, in May 2018, EPA's Office of Land and Emergency Management developed technical guidance for all its regions and programs for collecting, documenting, and managing

---

[57] According to 2018 EPA guidance, site boundaries identify the geographic extent of the site as a whole, including areas of contamination, and those boundaries change over time. Environmental Protection Agency, Office of Land and Emergency Management, *Collection and Documentation of General Descriptive Geospatial Site Data*, Version 3.4 (May 2018).

[58] Geographic standards include, for example, using the same geodetic datum, which, according to a NOAA website, uses a reference surface (such as sea level) to provide known locations to begin surveys and create maps. See https://www.ngs.noaa.gov/datums/index.shtml, accessed on July 26, 2019.

[59] Environmental Protection Agency, Office of Superfund Remediation and Technology Innovation, *Memo on Geospatial Superfund Site Data Definitions and Recommended Practices*, OLEM Directive 9200.2-191 (Nov. 29, 2017).

geographic information on Superfund sites, including their boundaries.[60] EPA officials told us that in 2019 and 2020, the agency plans to move toward recording site boundaries in a consistent format across regions and instituting procedures to validate and update them at least annually.

Table 1. Extent to Which EPA's Actions to Manage Risks to Human Health and the Environment from the Potential Impacts of Climate Change Effects at Nonfederal National Priorities List Sites Aligned with GAO's Essential Elements of Enterprise Risk Management

Essential elements	Extent to which EPA's actions aligned with essential elements
Aligning enterprise risk management process with goals and objectives	Not aligned
Identifying risks	Aligned
Assessing risks	Partially aligned
Responding to risks	Partially aligned
Monitoring risks	Aligned
Communicating about and reporting on risks	Aligned

Sources: GAO-17-63 and GAO analysis of Environmental Protection Agency (EPA) and stakeholder information. | GAO-20-73

However, EPA officials told us that there is no schedule in place for completing this effort and they are uncertain when they will complete it because of competing priorities. By developing a schedule for completing the standardization and improvement of the quality of the information on the boundaries of nonfederal NPL sites, EPA could more reasonably ensure that it would have quality information with which to fully identify nonfederal NPL sites that are located in areas that may be impacted by climate change effects.

---

60 Environmental Protection Agency, Office of Land and Emergency Management, *Collection and Documentation of General Descriptive Geospatial Site Data,* Version 3.4 (May 2018).

EPA Has Taken Some Actions to Manage Risks from the Potential Impacts of Climate Change Effects at Nonfederal NPL Sites

EPA's actions to manage risks from the potential impacts of climate change effects align with three of the six essential elements of enterprise risk management.[61] Specifically, for the six essential elements, EPA's actions do not align with one essential element, aligning its enterprise risk management process with goals and objectives; partially align with two essential elements, assessing risks and responding to risks; and align with three essential elements, identifying risks, monitoring risks, and communicating about and reporting on risks. Table 1 shows the alignment of EPA's actions with the essential elements of enterprise risk management.

Aligning Risk Management Process with Goals and Objectives

This essential element calls for agencies to align their risk management processes with the goals and objectives of the agency, but EPA has not taken action to clearly align its process for managing risks to human health and the environment from the potential impacts of climate change effects at nonfederal NPL sites with agency-wide goals and objectives. For example, the 2018 to 2022 EPA strategic plan does not include goals and objectives related to climate change or discuss strategies for addressing the impacts of climate change effects.[62] Moreover, neither the fiscal years 2018 to 2019 nor fiscal years 2020 to 2021 national program manager guidance for EPA's Office of Land and Emergency Management mentions climate

---

[61] We reported in 2016 that this enterprise risk management framework can help agencies assess threats that could affect the achievement of their goals: see GAO-17-63.

[62] Environmental Protection Agency, *Working Together: FY 2018-2022 EPA Strategic Plan*, EPA-190-R-18-003 (Washington, D.C.: February 2018).

change among its goals and priorities.[63] In contrast to the current strategic plan, the 2014 to 2018 EPA strategic plan included addressing climate change as one of four strategic goals and specifically discussed climate change as an external factor or emerging issue in the context of planned, current, and completed cleanups, including at nonfederal NPL sites. In addition, the fiscal years 2016 to 2017 national program manager guidance for the office that oversees the Superfund program listed climate change adaptation as one of four national areas of focus for the office.[64]

According to an EPA official, when the 2018 to 2022 strategic plan was drafted, senior agency officials were not aware of the potential risks to the Superfund program mission from the impacts of climate change effects. According to this official, senior EPA officials have expressed support for certain activities related to climate change, such as the work of the Cross-EPA Work Group on Climate Adaptation,[65] but have not issued related documents or policy statements.[66] Without clarifying how the agency's ongoing actions to manage these risks at nonfederal NPL sites align with current agency goals and objectives, EPA will not have reasonable assurance that senior officials will take an active role in supporting these actions, which would help EPA achieve its mission of protecting human health and the environment.

---

[63] Environmental Protection Agency, *Final FY 2018-2019 Office of Land and Emergency Management National Program Manager Guidance*, 540B17001 (Sept. 29, 2017), and *Final FY 2019-2020 Office of Land and Emergency Management National Program Manager Guidance,* 500B19002 (June 7, 2019). The national program manager guidance communicates operational planning priorities, strategies, and key activities for advancing the agency's strategic plan.

[64] Environmental Protection Agency, *Fiscal Year 2014-2018 EPA Strategic Plan*, EPA-190-R14-006 (Washington, D.C.: Apr. 10, 2014), and *Office of Solid Waste and Emergency Response FY 2016-2017 National Program Manager Guidance*, 530R15001 (Apr. 28, 2015).

[65] The Cross-EPA Work Group on Climate Adaptation shares information among regional and program offices to support the integration of climate adaptation planning into EPA's programs and operations, including cleanup at nonfederal NPL sites.

[66] In 2011 and 2014, EPA administrators issued policy statements indicating that EPA would take actions to address the impacts of climate change effects. The two most recent EPA administrators have not made policy statements about climate change, according to an EPA official.

Identifying Risks

EPA's actions to identify risks to human health and the environment from the potential impacts of climate change effects at nonfederal NPL sites align with this essential element of enterprise risk management. Specifically, EPA identified climate change effects that may impact nonfederal NPL sites—and pose risks to human health and the environment—in studies and climate change adaptation and implementation plans. For example, in a 2012 study of adaptation of Superfund remediation to climate change, EPA identified eight climate change effects that may impact certain NPL site remedies: flooding, sea level rise, extreme storms, large snowfall, wildfires, drought, extreme heat, and landslides.[67] In 2014, EPA issued an agency-wide climate change adaptation plan, which identified climate change effects that may impact NPL sites.[68] The same year, EPA issued a climate change adaptation implementation plan for the office that oversees the Superfund program that identified nine climate change effects that may impact NPL sites.[69]

Each of the 10 EPA regional offices identified relevant regional climate change effects in their 2014 climate change adaptation implementation plans.[70] For example, the Region 3 plan states that increased flooding and sea level rise may increase risks of releases of contaminants, salt water intrusion may impact the performance of

[67] Environmental Protection Agency, *Adaptation of Superfund Remediation to Climate Change* (Washington, D.C.: February 2012). According to an EPA headquarters official, the agency does not plan to update this 2012 study, because the data upon which its conclusions are based remain valid.

[68] Environmental Protection Agency, *U.S. Environmental Protection Agency Climate Change Adaptation Plan*, EPA 100-K-14-001 (June 2014).

[69] Environmental Protection Agency, *Office of Solid Waste and Emergency Response Climate Change Adaptation Implementation Plan* (June 2014).

[70] In 2014, EPA developed an agency-wide climate change adaptation plan and climate change adaptation implementation plans for each EPA program office and the 10 EPA regions in response to Executive Order 13653, issued in 2013, which directed each federal agency to evaluate climate change risks and vulnerabilities to the agency's mission and operations in both the short and long term. Exec. Order No. 13653, 78 Fed. Reg. 66819 (Nov. 6, 2013). The adaptation plan stated that EPA will take the actions necessary to ensure that it continues to fulfill its mission of protecting human health and the environment even as the climate changes. Executive Order 13653 was revoked in 2017 by Executive Order 13783. Exec. Order No. 13783, 82 Fed. Reg. 16093 (Mar. 31, 2017).

remedies, and increased temperatures may impact vegetation that prevents erosion. Additionally, five regional offices have conducted or are conducting additional screening-level studies to identify which climate change effects, if any, may impact each of the NPL sites in these regions.[71] For example, Region 10 conducted a study in 2015 that identified, among other effects, sea level rise and wildfires as potential climate change effects that may impact NPL sites in the region.

Assessing Risks

EPA's actions to assess risks to human health and the environment from the potential impacts of climate change effects at nonfederal NPL sites partially align with this essential element. In a 2012 study of adaptation of Superfund remediation to climate change, EPA assessed the impacts of eight climate change effects on certain remedies to determine the risk they presented to the agency's mission to protect human health and the environment.[72] EPA issued climate change adaptation implementation plans for the office that oversees the Superfund program and all regions, as described above, which assessed potential impacts of climate change effects. In addition, five EPA regional offices assessed or are assessing potential impacts of climate change effects on NPL sites in their regions as a whole, and one of these regions assessed both the impacts and likelihood of climate change effects, consistent with this essential element. Specifically, Region 4 identified the sites most likely to face major climate change risks and then examined these sites in greater detail.[73] Additionally, Region 3 has developed a mapping tool on climate change vulnerability that provides site-level assessments of sea level rise, among other potential impacts.

---

[71] These are regions 1, 3, 4, 6, and 10.

[72] Environmental Protection Agency, *Adaptation of Superfund Remediation to Climate Change* (Washington, D.C.: February 2012).

[73] Environmental Protection Agency and Industrial Economics, Inc., *Methodology for Ranking the Climate Change Vulnerability of NPL Sites* (Mar. 19, 2015).

EPA provides training and direction to remedial project managers—the lead EPA officials at nonfederal NPL sites—on conducting site-level risk assessments that incorporate information on potential impacts of climate change effects. Since 2014, EPA has offered optional training to remedial project managers and others on integrating climate change into the Superfund cleanup process. From 2013 through 2015, EPA issued fact sheets as guidance for assessing the potential impacts of climate change effects for three types of remedies.[74] According to EPA officials, these fact sheets constitute the direction that EPA provides to remedial project managers on assessing risks from climate change effects. EPA plans to update these fact sheets in 2019 and is also in the process of developing a compilation of resources for assessing potential flood risks in coastal areas to inform cleanup and reuse decision-making, according to an EPA official. In addition, EPA provides resources on climate change on the Superfund program website, such as links to tools and data on drought and coastal flooding.[75] EPA also offers technical assistance on incorporating climate change information into risk assessments to remedial project managers through groups such as the Contaminated Sediments Technical Advisory Group and the Cross-EPA Work Group on Climate Adaptation.[76]

EPA officials in four regions provided us with site-specific examples of how they used climate change information to assess risks from the

---

[74] Environmental Protection Agency, *Climate Change Adaptation Technical Fact Sheet: Groundwater Remediation Systems*, EPA 542-F-13-004 (December 2013); *Climate Change Adaptation Technical Fact Sheet: Landfills and Containment as an Element of Site Remediation*, EPA 524-F-14-001 (May 2014); and *Climate Change Adaptation Technical Fact Sheet: Contaminated Sediment Remedies*, EPA 542-F-15-009 (April 2015).

[75] Environmental Protection Agency, *Superfund Climate Resilience: Vulnerability Assessment*, accessed June 27, 2019, https://www.epa.gov/superfund/superfund-climate- resilience-vulnerability-assessment. In addition, EPA has compiled a more general collection of resources about climate change adaptation for local government officials on the Climate Change Adaptation Resource Center (ARC-X) website, including tools, case studies, and training materials. As of July 2019, the site included one case study on climate change adaptation at a nonfederal NPL site. According to an EPA official, the agency is working to develop additional Superfund case studies. Also, in June 2019, Region 10 launched a Climate Adaptation Action Forum that will share information on climate change issues and is open to all EPA staff, according to an EPA official.

[76] The Contaminated Sediments Technical Advisory Group reviews cleanup decisions at certain sediment sites and has recommended that EPA remedial project managers require the use of the most recent climate information in characterization data and models, according to an EPA official.

potential impacts of climate change effects, but officials from other regions stated that they have not always integrated climate change information into their risk assessments. For example, according to a record of decision for the site, EPA Region 2 incorporated the potential for increased storm flow intensities into the model of the Passaic River used in the remedial investigation and feasibility study at the Diamond Alkali site in Newark, New Jersey.[77] Conversely, officials in six regions told us that they have not used climate change projections for flooding or rainfall in site-level risk assessments. In addition, officials in Region 6 told us that they do not incorporate potential impacts of climate change effects or changes in the frequency of natural disasters into their assessments.

EPA officials have not consistently incorporated climate change information into their assessment of site-level risks because they do not always have the climate data they need to do so, according to our review of EPA documents and interviews with EPA officials and stakeholders. For example, officials in three regions told us that they have not used rainfall or flood projections because the data are not available or they were unsure which data to use. In addition, in the record of decision for the Diamond Alkali site in New Jersey, Region 2 officials stated that they did not integrate sea level rise information into their storm flow modeling for the Passaic River at the site because of the uncertainty in expected future sea level rise values, especially at the regional and local levels. We reported on similar challenges with climate data in our 2015 report on climate information, which found that existing federal efforts do not fully meet the climate information needs of federal, state, local, and private sector decision makers, and we made a related recommendation in that report.[78]

---

[77] The soil, groundwater, and sediment in rivers and bays at the Diamond Alkali Superfund site are contaminated with dioxins, pesticides, polychlorinated biphenyls, and other hazardous substances from past manufacturing activities.

[78] GAO, *Climate Information: A National System Could Help Federal, State, Local, and Private Sector Decision Makers Use Climate Information*, GAO-16-37 (Washington, D.C.: Nov. 23, 2015). In that report, we recommended that the Executive Office of the President designate a federal entity to develop and periodically update a set of authoritative climate change observations and projections for use in federal decision-making and create a national climate information system with defined roles for federal agencies and nonfederal entities. The Executive Office of the President did not respond to our recommendations, and as of March 2019, has not implemented the recommendations. See GAO-19-157SP.

Further, current EPA practice for assessing risks at NPL sites does not always include consideration of climate change, according to agency documents we reviewed and officials from three regions and a stakeholder we interviewed. EPA's climate change adaptation plan noted that EPA and its partners will need to alter their standard practices—such as their standard methods for estimating the frequency of floods or runoff of pollutants into rivers—to account for a continuously changing climate. The Region 4 climate change adaptation implementation plan, for instance, noted that preliminary assessments and site investigations are typically based on historic information, not future projections and therefore may not fully address risks. Officials in two regions told us that they do not have direction on how to alter their practices to account for climate change. For example, officials in Region 2 said they do not have instructions that identify a particular set of expectations, data, or maps that they should use when considering future risks from flooding. Officials in Region 5 told us that they do not have any formal direction on how to address risks from climate change and are waiting for EPA headquarters to provide information on how to do so.

According to EPA documents and a headquarters official, EPA believes that its existing direction, including general guidance on conducting risk assessments and the fact sheets for assessing potential impacts of climate change effects for three types of remedies, discussed above, provide a robust structure for considering such impacts. However, without providing direction to remedial project managers on how to integrate information on the potential impacts of climate change effects into site- level risk assessments at nonfederal NPL sites across all regions and types of remedies, EPA cannot ensure that remedies will protect human health and the environment in the long term.

Responding to Risks

EPA's actions to respond to risks that potential impacts of climate change effects may pose to human health and the environment at nonfederal NPL sites partially align with this essential element. In two national studies EPA conducted in 2012 and 2017, EPA examined potential impacts of some climate change effects on selected remedies at NPL sites, including nonfederal NPL sites, and generally found that it has taken actions to respond to risks through its existing cleanup processes. In 2012, as noted above, EPA studied the vulnerability of selected remedies to some climate change effects and found that existing processes—such as EPA's Five-Year Review and operation and maintenance—could adequately address the potential impacts of climate change effects. In addition, EPA studied the impacts of three hurricanes in 2017 on sites with selected remedies in place, including nonfederal NPL sites, and found that the agency has generally taken resiliency measures to respond to risks at these sites.[79]

EPA also provided guidance and training to remedial project managers on responding to risks to human health and the environment from the potential impacts of climate change effects and recently added requirements for certain potential site contractors to describe their capacity to respond to such risks. EPA provided guidance in its fact sheets on integrating climate change information into risk response decisions at nonfederal NPL sites and optional training on integrating climate change into the Superfund cleanup process. In addition, EPA provided relevant information and resources for EPA officials on resiliency measures on the agency website. In 2016, EPA issued performance work statements to potential contractors for environmental services and operations and for remediation environmental services that required contractors to describe their ability to conduct climate change vulnerability analyses and adaptation, as needed, to ensure the resiliency of remedies to climate change impacts.[80] According to an EPA headquarters official, EPA is currently working on developing technical guidance on how remedial

---

[79] Environmental Protection Agency, *Evaluation of Remedy Resilience at Superfund NPL and SAA Sites* (August 2018).

[80] Environmental Protection Agency, *Remediation Environmental Services Contract Performance Work Statement*, SOL-HQ-14-00023, Attachment 1 (Sept. 16, 2016) and *Environmental Services and Operations Performance Work Statement*, SOL-R1-14-00003, Attachment 1 (Feb. 26, 2016).

project managers can integrate requests for climate change– related analysis into their task orders for contractors.

With respect to site-level responses, EPA officials from three regions provided us with examples of site decision documents that described how climate change information will be incorporated into remedy selection and design. For example, the record of decision for the Portland Harbor site in Oregon states that a containment cap will be constructed to withstand more frequent floods with higher peak flows more common with climate change. Officials from Region 3 told us that they take into account a number of factors, including climate change impacts, if any, when they design and select site remedies.

However, according to our interviews with regional officials, they have not consistently integrated climate change information into remedy selection and design. For example, officials from two regions stated that they are not aware of any remedial project managers in their regions who are taking action at nonfederal NPL sites to respond to climate change or consider future conditions. EPA officials have not consistently taken the potential impacts of climate change effects into account in site-level risk response decision making because they do not always have sufficient direction to do so, according to our interviews with EPA officials. EPA officials from three regions told us that they are unsure how to translate data on potential impacts of climate change effects into the design of remedies. For example, officials from Region 10 told us that EPA does not have direction for remedial project managers on how to integrate response to climate change impacts into remedial design. These officials noted that it is up to remedial project managers to be aware of this issue and it is done on an ad hoc basis. Further, EPA headquarters officials who review proposed remedies told us that additional guidance from EPA on managing the risks from potential impacts of climate change effects would be useful.

According to EPA documents and another EPA headquarters official, EPA has determined that existing direction—guidance and processes— for risk response provide a robust structure to integrate climate change information into remedy selection and design. However, without providing direction for remedial project managers on how to integrate information on potential impacts of climate change effects into site-level risk response decision making at nonfederal NPL sites, EPA cannot ensure that remedies will protect human health and the environment in the long term.

Monitoring Risks

EPA's actions to monitor risks to human health and the environment from the potential impacts of climate change effects at nonfederal NPL sites through its Five-Year Review process align with this essential element. In 2016, EPA introduced a new recommended template for the Five-Year Review that includes a section for officials to document their consideration of whether any newly available information related to climate change may call into question a remedy's protectiveness.[81]

Officials in three regions told us they use the Five-Year Review process to identify and evaluate newly available information on climate change effects that may impact nonfederal NPL sites. For example, in the 2014 Five-Year Review report for the Publicker Industries site in Pennsylvania, Region 3 considered newly available information on projected sea level rise in the region to determine if those projections called into question the protectiveness of the existing remedies at the site. Officials in that region told us that they rely on their biological and technical assistance group to identify any new relevant climate change data to incorporate into their Five-Year Reviews. Region 7 officials also told us that they assess any potential changes in future conditions, especially flooding, during the Five-Year Review process. Officials from two other regions told us that they monitor changes in site conditions that may be related to climate change during the Five-Year Review process. For example, Region 2 officials developed additional guidance to help remedial project managers and site project teams consider changes in site

[81] Environmental Protection Agency, *Five-Year Review Recommended Template*, OLEM 9200.0-89 (Washington, D.C.: Jan. 20, 2016). According to an EPA memorandum, the template provides officials with an approach for preparing the Five-Year Review reports in a manner intended to promote national consistency, to reduce nonessential information, and to decrease repetitiveness in the reports.

conditions related to climate change in the Five-Year Review process. Region 6 officials told us that during the Five-Year Review process, they take into account any current flood hazard information from FEMA as well as current sea levels, but they do not monitor projections about sea level rise.

Communicating and Reporting on Risks

EPA's actions to communicate about and report on risks to human health and the environment from the potential impacts of climate change effects at nonfederal NPL sites align with this essential element. For example, as described above, EPA reported on the potential impacts of climate change effects—which may pose risks to human health and the environment—on NPL sites in its 2014 national climate change adaptation plan and the climate change adaptation implementation plans for the office that oversees the Superfund program and all regions. In addition, publicly available site-level documents, such as the records of decision described above, may include information on risks from climate change and EPA's actions to manage these risks. EPA officials may also communicate this information in response to questions from the public. EPA officials from four regions told us that they have not received many direct questions on risks from climate change from the public. However, members of the public can comment on climate change risks through EPA's existing public engagement mechanisms,[82] and some have done so. For example, EPA officials in Region 7 received questions on the draft record of decision for the West Lake Landfill site in Missouri during the public comment period and responded to those questions in the final version of the document, describing how they addressed risks of increased flooding from climate change in the remedy selection processes.

[82] According to EPA, EPA's Superfund Community Involvement Program is to provide individuals affected by hazardous waste sites with information and opportunities to participate in the decisions that affect the Superfund sites in their communities. Further, the program seeks to ensure that each community has a voice during all phases of the Superfund cleanup process.

EPA has also communicated with stakeholders and the public on risks to human health and the environment from the potential impacts of climate change effects in other ways. For example, officials from Region 10 convened a workshop in 2017 to discuss climate change impacts on sediment cleanup and upland source control for the Lower Duwamish Waterway site in Washington with other federal agencies, state and local officials, universities, companies, and community groups. In addition, EPA provides an online mapping tool that can help members of the public identify sites located in areas that would be impacted by up to 6 feet of sea level rise or in flood hazard areas as determined by FEMA.[83]

EPA Recognizes Various Challenges in Managing Risks from the Potential Impacts of Climate Change Effects at Nonfederal NPL Sites

EPA recognizes institutional, resource, and technical challenges in managing risks to human health and the environment from the potential impacts of climate change effects at nonfederal NPL sites, according to agency and other documents that we reviewed and EPA officials and stakeholders we interviewed.

Institutional Challenges

According to agency and other documents we reviewed and officials and stakeholders that we interviewed, EPA faces institutional challenges in managing risks to human health and the environment from the potential impacts of climate change effects. As discussed above, officials from three regions told us that they do not have the direction they need to manage

[83] The Cleanups in My Community online tool maps and lists hazardous waste cleanup locations and grant areas, such as nonfederal NPL sites. EPA also integrates other federal data into the tool, including sea level rise scenarios and FEMA flood hazard areas. Accessed April 8, 2019, at https://www.epa.gov/cleanups/cleanups-my-community.

these risks. For example, EPA officials in Region 2 told us that during Five-Year Reviews, engineers may analyze several different maps on flooding potential and must use their professional judgment to determine how resilient to design the remedy, because there is no standard guidance on how to do so. Further, EPA officials in two regions and stakeholders we interviewed stated that it may not be clear whether EPA could require PRPs to consider climate change impacts in the cleanup process. However, according to EPA headquarters officials, considering climate change is consistent with the National Contingency Plan and the CERCLA criterion that requires officials to consider the long- term effectiveness of remedies when evaluating cleanup alternatives.

Another institutional challenge that EPA faces is that its ability to manage these risks may depend on actions of other entities that are outside of EPA's control, according to EPA documents we reviewed and EPA officials we interviewed. For example, EPA officials from Region 1 told us that they are not certain whether a hurricane barrier built by the U.S. Army Corps of Engineers that protects the New Bedford Harbor site in Massachusetts is designed to withstand future storms. Managing risks may also require internal coordination within EPA, which presents another challenge. For example, an EPA headquarters official told us that it can be challenging for regional Superfund program staff to connect with EPA experts on climate change, who may be in different program offices. In April 2019, EPA restructured its regional offices, consolidating cross-cutting issue areas in the immediate office of each Regional Administrator and Deputy Regional Administrator. Although it is too early to evaluate the effect of this restructuring, EPA headquarters officials told us that the restructuring may help address this challenge. Furthermore, EPA officials from three regions told us that they face challenges related to the sensitive nature of climate change. For example, officials in Region 6 told us that when they engaged with the local community during the decision making process for the San Jacinto River Waste Pits site in Texas, they avoided using the term climate change because of concerns that the charged term would alienate some community members.

Resource Challenges

Documents from four EPA regions and headquarters officials and officials from three regions we interviewed stated that insufficient or changing resources—specifically funding and staffing—makes managing risks to human health and the environment from the potential impacts of climate change effects challenging for EPA. For example, according to two regional climate change adaptation implementation plans and EPA officials, assessing these risks may require more resources than assessing risks based on current or past conditions. In addition, designing or modifying existing remedies to respond to these risks could increase costs, according to EPA documents we reviewed and EPA officials we interviewed.

EPA officials from three regions told us that staffing constraints can make it difficult to manage risks. For example, EPA officials from Region 9 told us that the need for remedial project managers to respond to other emergencies, such as overseeing hazardous materials removal after fires, means that they have less time to oversee cleanup of nonfederal NPL sites.[84] Officials from Region 10 told us that they had a climate change advisor who helped integrate climate change into all aspects of the region's work, but that person retired, and the region was unable to fill the position because of resource constraints. As noted above, according to an EPA headquarters official, EPA's recent restructuring of its regional offices may help address this challenge.

---

[84] EPA's 2019 guidance on planning for natural disaster debris reported that the amount of debris generated by natural disasters will likely increase in the future as a result of climate change. The document also notes that disaster debris–related consequences of major natural disasters may include a greater risk of releases from Superfund sites and that local decision makers should plan accordingly. Environmental Protection Agency, *Planning for Natural Disaster Debris*, EPA 530-F-19-003 (April 2019).

Technical Challenges

EPA faces technical challenges in managing risks to human health and the environment from the potential impacts of climate change effects in terms of available expertise and data, according to documents we reviewed and EPA regional officials we interviewed. In its 2014 agency- wide climate change adaptation plan, EPA reported that site vulnerabilities may be difficult to assess because of limited scientific understanding. EPA officials told us that they need additional expertise and training to better manage risks. For example, an EPA official in Region 2 told us that it would be useful to have training on assessing risks for projects located in floodplains. As noted above, EPA has developed training for officials on managing risks from climate change, such as a course on building resilient Superfund remedies that EPA offered at the annual National Association of Remedial Project Managers meeting in August 2019. The course's focus is to help remedial project managers incorporate consideration of adaptation and build resilience into Superfund remedies at extreme weather event–impacted sites, according to the course agenda.

According to EPA documents and EPA officials from two regions, appropriate climate change data may not be available to inform assessments that help manage risk. For example, the Region 4 study of the vulnerability of NPL sites stated that climate model projections of temperature and precipitation patterns are not available at a spatial resolution that is useful for assessing vulnerabilities at the site level. In Region 6, officials told us that when the U.S. Army Corps of Engineers modeled flooding for the San Jacinto River Waste Pits site in Texas, it had to rely on past flooding data because the only information available was on historical storms.[85]

---

[85] U.S. Army Corps of Engineers, Engineer Research and Development Center, *Modeling the Impact of Hurricane Ike and the October 1994 San Jacinto River Flood on the San Jacinto Waste Pits Cap* (September 2017).

In addition, the level of uncertainty inherent in climate change data may make it challenging for EPA to incorporate that information into risk management processes, according to agency documents we reviewed and some agency officials we interviewed. As noted above, we made recommendations to address similar challenges with climate data in a prior report.[86]

CONCLUSION

Climate change may result in more frequent or intense extreme events, such as flooding, storm surge, and wildfires, among other effects, which could damage remedies at nonfederal NPL sites and lead to releases of contaminants that could pose risks to human health and the environment. Our analysis of EPA, FEMA, NOAA, and U.S. Forest Service data has shown that more than half of nonfederal NPL sites—based on a point coordinate with a 0.2-mile radius as a proxy for the site boundaries—are located in areas that may be impacted by selected climate change effects. To help ensure the long-term protectiveness of remedies, it is important for EPA to understand how climate change effects may impact nonfederal NPL sites, and the agency has taken steps to do this. However, EPA does not have quality information on the precise boundaries of nonfederal NPL sites, which could make it difficult to determine the nonfederal sites located in areas that may be impacted by climate change effects. The agency has taken initial steps to develop this information but does not have a schedule in place for completing this effort.

EPA has taken actions to manage risks to human health and the environment from the potential impacts of climate change effects at nonfederal NPL sites. These actions align with three of the six essential elements of enterprise risk management.

---

[86] See GAO-16-37.

However, EPA has not clarified how its actions to manage risks from these effects at nonfederal NPL sites align with current agency goals and objectives, which could limit its senior officials' ability to manage these risks. Further, EPA officials do not always have direction to ensure that they consistently integrate climate change information into site-level risk assessments and risk response decisions, according to EPA documents and officials. Without providing such direction for remedial project managers, EPA cannot ensure that remedies at nonfederal NPL sites will protect human health and the environment in the long term.

RECOMMENDATIONS FOR EXECUTIVE ACTION

We are making the following four recommendations to EPA:

- The Director of the Office of Superfund Remediation and Technology Innovation should establish a schedule for standardizing and improving information on the boundaries of nonfederal NPL sites. (Recommendation 1)
- The Administrator of EPA should clarify how EPA's actions to manage risks to human health and the environment from the potential impacts of climate change effects at nonfederal NPL sites align with the agency's current goals and objectives. (Recommendation 2)
- The Director of the Office of Superfund Remediation and Technology Innovation should provide direction on how to integrate information on the potential impacts of climate change effects into risk assessments at nonfederal NPL sites. (Recommendation 3)

- The Director of the Office of Superfund Remediation and Technology Innovation should provide direction on how to integrate information on the potential impacts of climate change effects into risk response decisions at nonfederal NPL sites. (Recommendation 4)

AGENCY COMMENTS AND OUR EVALUATION

We provided a draft of this chapter to EPA for its review and comments. In its comments, reproduced in appendix II, EPA stated that it recognizes the importance of ensuring Superfund sites cleanups are resilient in the face of existing risks and extreme weather events. EPA added that it has taken actions to include vulnerability analyses and adaptation planning in its Superfund activities. We acknowledge that EPA has taken some action to manage risks. However, EPA has not clarified how its risk-related actions align with agency goals and objectives. Further, it has not provided direction to ensure that officials consistently integrate climate change information into site-level risk assessments and risk response decisions.

Regarding our recommendations, EPA agreed with one and disagreed with the other three. We continue to believe that all recommendations are warranted.

In response to our recommendation that the Director of the Office of Superfund Remediation and Technology Innovation establish a schedule for standardizing and improving information on the boundaries of nonfederal NPL sites, EPA noted that it agrees with our finding and acknowledges a lack of consistent standards to identify site boundaries at the national level. According to EPA, it has taken initial steps to develop an approach to standardize and improve information on nonfederal NPL site boundaries.

EPA stated that it expects to establish a schedule for this effort by the second quarter of fiscal year 2020, with the aim to have collected an initial set of site boundaries for all NPL sites by the fourth quarter of fiscal year 2021.

In response to our recommendation that EPA clarify how its actions to manage risks to human health and the environment from the potential impacts of climate change effects at nonfederal NPL sites align with the agency's current goals and objectives, EPA said that it believes managing risks from exposure to contaminants in the environment is integral to EPA's current strategic goal 1.3, Revitalize Land and Prevent Contamination. We agree that protectiveness is a key part of strategic objective 1.3. However, this strategic objective does not include any measures related to climate change or discuss strategies for addressing the impacts of climate change effects. An essential element of enterprise risk management is to align risk management processes with goals and objectives. Consequently, we believe that our recommendation is still warranted.

In response to our recommendations that the Director of the Office of Superfund Remediation and Technology Innovation provide direction on how to integrate information on the potential impacts of climate change effects into risk assessments and risk response decisions at nonfederal NPL sites, EPA said that it strongly believes the Superfund program's existing processes and resources adequately ensure that risks and any effects of severe weather events are woven into risk assessments and risk response decisions at nonfederal NPL sites. However, as we noted in our report, EPA's current direction does not address all types of cleanup actions or climate change effects. Further, EPA officials from some regions told us that current EPA practice for assessing risks at NPL sites does not always include consideration of climate change and that they have not consistently integrated climate change information into site- specific remedy selection and design. EPA noted that it may issue a memorandum to reinforce the tools and resources available to NPL site teams and would determine whether to issue this memorandum by the end of January 2020.

EPA also provided technical comments, which we incorporated as appropriate.

As agreed with your offices, unless you publicly announce the contents of this chapter earlier, we plan no further distribution until 30 days from the report date. At that time, we will send copies to the appropriate congressional committees, the Administrator of the Environmental Protection Agency, and other interested parties.

J. Alfredo Gómez
Director, Natural Resources and Environment

List of Requesters

The Honorable Thomas R. Carper
Ranking Member
Committee on Environment and Public Works
United States Senate

The Honorable Sheldon Whitehouse
Ranking Member
Subcommittee on Clean Air and Nuclear Safety
Committee on Environment and Public Works
United States Senate

The Honorable Tammy Duckworth
Ranking Member
Subcommittee on Fisheries, Water, and Wildlife
Committee on Environment and Public Works
United States Senate

The Honorable Cory Booker
Ranking Member
Subcommittee on Superfund, Waste Management, and Regulatory Oversight
Committee on Environment and Public Works
United States Senate

The Honorable Benjamin L. Cardin
Ranking Member
Subcommittee on Transportation and Infrastructure
Committee on Environment and Public Works
United States Senate

The Honorable Betty McCollum
Chair
Subcommittee on Interior, Environment, and Related Agencies
Committee on Appropriations
House of Representatives

The Honorable Kirsten E. Gillibrand
United States Senate

The Honorable Kamala Harris
United States Senate

The Honorable Edward J. Markey
United States Senate

The Honorable Jeffrey A. Merkley
United States Senate

The Honorable Bernard Sanders
United States Senate

APPENDIX I: OBJECTIVES, SCOPE, AND METHODOLOGY

This chapter examines (1) what available federal data suggest about the number of nonfederal National Priorities List (NPL) sites that are located in areas that may be impacted by selected climate change effects; (2) the extent to which the Environmental Protection Agency (EPA) has managed risks to human health and the environment from the potential impacts of climate change effects at nonfederal NPL sites; and (3) the challenges, if any, EPA faces in managing these risks.

To determine what available federal data suggest about the number of nonfederal NPL sites that are located in areas that may be impacted by selected climate change effects, we reviewed the Fourth National Climate Assessment (NCA) to identify potential climate change effects.[87] Based on our review of the NCA, we identified the following potential climate change effects: sea level rise, which may lead to increased frequency and extent of extreme flooding from coastal storms; greater frequency and magnitude of drought; increased intensity and frequency of heavy precipitation events, which may lead to increased local flooding; salt water intrusion; increased incidence of large wildfires; increased frequency and intensity of extreme high temperatures and sustained increases in average temperatures; decreased permafrost; and increased intensity—including higher wind speeds and precipitation rates—and frequency of very intense hurricanes and typhoons. We reviewed EPA documents (such as EPA's climate change adaptation implementation plans) to identify potential climate change effects that may impact nonfederal NPL sites and interviewed EPA officials.

Through a review of federal agencies' documents and databases and interviews with officials about their data and research on these effects, we identified available national federal data sets on three current hazards: flooding, storm surge, and wildfires—which the NCA reports will be

---

[87] U.S. Global Change Research Program, *Climate Science Special Report: Fourth National Climate Assessment, Volume I* (Washington, D.C.: 2017) and *Impacts, Risks, and Adaptation in the United States: Fourth National Climate Assessment, Volume II* (Washington, D.C.: 2018).

exacerbated by climate change—from the Federal Emergency Management Agency (FEMA), the National Oceanic and Atmospheric Administration (NOAA), and the U.S. Forest Service.[88] We also identified data on sea level rise from NOAA.

In this chapter, we refer to (1) flooding, (2) storm surge, (3) wildfires, and (4) sea level rise as potential climate change effects. We used the most recently available data for each of these climate change effects; these data do not provide estimates of the projected changes in the future. To the extent that data were available, we analyzed a range of these potential climate change effects. For example, we used the maximum extent of storm surge from Category 1 hurricanes as well as Category 4 or 5 hurricanes, the highest possible categories, as modeled by NOAA. We focused on a range because, for three of the four effects, we had data on current hazards, which may become more intense and frequent in the future, according to the NCA. Additionally, the Comprehensive Environmental Response, Compensation, and Liability Act of 1980 (CERCLA) directs EPA to give preference to remedies that would result in the permanent and significant decrease in toxicity, mobility, or volume of the contamination. According to EPA officials, remedies at nonfederal NPL sites may have to be operational indefinitely, during which time the potential effects of climate change may become more extreme.

The range of estimates we provide in our report is as follows:

- For flooding, we used data from FEMA's National Flood Hazard Layer as of October 2018. FEMA identifies a variety of flood hazards, and for reporting purposes, we grouped flood hazard zones into four categories: (1) 1 percent or higher annual chance of flooding,[89] (2) 0.2 percent or higher annual chance of flooding or other flood hazards,[90] (3) unknown flood hazards,[91] and (4) minimal flood hazard.

---

[88] We did not include forward-looking data or projections in our analysis on flooding, storm surge, and wildfire because these data are not available at a national level.

[89] This includes zones A, A99, AE, AH, AO, V, VE, and Open Water.

[90] This category includes hazards FEMA categorizes in zone X (excluding minimal flood hazard).

- For storm surge, we used data from NOAA's model on Sea, Lake and Overland Surges from Hurricanes as of November 2018 for Category 1 and Category 4 or 5 hurricanes.
- For wildfire, we used data from the U.S. Forest Service's 2018 wildfire hazard potential map, which the U.S. Forest Service released in July 2018. We used areas with high or very high wildfire hazard potential in our analysis. The U.S. Forest Service based the 2018 map on wildfire likelihood and intensity data from 2016, spatial fuels and vegetation data from 2012, and point locations of past fire occurrence from 1992 to 2013.
- For sea level rise, we used NOAA data, last updated in September 2018. We downloaded inundation data on 0, 1, 3, and 8 feet of sea level rise and "not mapped" areas.

We obtained data from EPA's Superfund Enterprise Management System on the location and other characteristics of active and deleted nonfederal NPL sites. In our analysis, we used a 0.2-mile radius around the primary geographic coordinate point of each nonfederal NPL site, which may not accurately represent their actual areas because the sites vary in size and shape.[92] The EPA data we used in our analysis on the location of nonfederal NPL sites are current as of March 2019. We also obtained EPA data on contaminants and types of remedies that are current as of fiscal year 2014 to determine the number of contaminants EPA has identified in nonfederal NPL sites. We did not conduct further site-specific analyses, such as those related to the extent of contamination and location of remedies. We reviewed documents from the Agency for Toxic Substances and Disease Registry on the health effects of hazardous substances in nonfederal NPL sites and interviewed officials from that agency.

---

[91] Unknown includes zones D, NP, missing values, area not included, and no data. In addition, we considered data that we could not analyze using our mapping software, such as those available in paper-based maps, as "unknown flood hazard."

[92] In a 2018 study, EPA used a 0.2-mile radius to approximate the size of NPL sites. In this study, EPA noted that it used additional information to adjust this radius for some NPL sites. We did not make such adjustments because it would have required site-specific analysis, which was outside the scope of our review. See Environmental Protection Agency, *Evaluation of Remedy Resilience at Superfund NPL and SAA Sites* (August 2018).

To analyze whether nonfederal NPL sites are located in areas that may be impacted by flooding, we used ArcGIS mapping software to intersect the area of a 0.2-mile radius around the primary coordinate of the sites with the categories we defined from the National Flood Hazard Layer. If a site overlapped with areas in more than one of the four reporting groups, we categorized the site in the group representing the highest flood hazard. For the purposes of our report, we considered the highest flood hazard to be, in descending order, 1 percent or higher annual chance of flooding, other flood hazards (including 0.2 percent or higher annual chance of flooding), unknown flood hazard or no data, and minimal flood hazard. To analyze whether nonfederal NPL sites are located in areas that may be impacted by storm surge, wildfires, and sea level rise, we used MapInfo mapping software to intersect the area of a 0.2-mile radius around the primary coordinates of sites with each of these layers. Overlap indicates that a site is potentially in an area that may be impacted.

To assess the reliability of FEMA's National Flood Hazard Layer, we reviewed FEMA's methodology, guidelines, and standards; interviewed FEMA officials to assess the timeliness and accuracy of the data as well as any limitations of the data; conducted data testing to check for missing data and inconsistencies; and reviewed internal controls. We also reviewed a prior GAO report on the methodology FEMA uses to map flood hazards.[93] To assess the reliability of NOAA's data on Sea, Lake and Overland Surges from Hurricanes, we reviewed NOAA's methodology for developing the model, interviewed NOAA officials to assess the timeliness and accuracy of the data as well as any limitations of the data, and reviewed internal controls. To assess the reliability of the U.S. Forest Service's wildfire hazard potential data, we reviewed the agency's documentation of the methodology, interviewed U.S. Forest Service officials to assess the timeliness and accuracy of the data as well as any limitations of the data, and reviewed internal controls. We also reviewed

---

[93] GAO, *FEMA Flood Maps: Some Standards and Processes in Place to Promote Map Accuracy and Outreach, but Opportunities Exist to Address Implementation Challenges*, GAO-11-17 (Washington, D.C.: Dec. 2, 2010).

our past reports that cited the 2014 versions of these data.[94] To assess the reliability of NOAA's data on sea level rise, we reviewed the methodology NOAA used for developing the model, interviewed NOAA officials to assess the timeliness and accuracy of the data as well as any limitations of the data, and reviewed internal controls.

To assess the reliability of EPA's data, we reviewed agency manuals and data dictionaries to understand data elements, interviewed EPA officials to assess the timeliness and accuracy of the data and related internal controls, conducted data testing, discussed inaccuracies with EPA officials; and obtained corrected data. For example, we compared the zip code of each nonfederal NPL site to its coordinate to check the accuracy of site locations. We shared potential errors with EPA officials, who corrected the coordinates of six sites. As a result of the steps described above, we found data from EPA, FEMA, NOAA, and the U.S. Forest Service to be sufficiently reliable for our purposes.

To determine the extent to which EPA has managed risks to human health and the environment from the potential impacts of climate change effects on nonfederal NPL sites, we examined relevant provisions in CERCLA, EPA's implementing regulations, and executive orders. We also reviewed EPA documents, including climate change adaptation and implementation plans; vulnerability studies; training materials; and site-specific documents, our prior work, and relevant documents from other organizations, such as the National Research Council. We identified these documents by conducting a search of (1) websites of relevant agencies and organizations and (2) article databases. We also reviewed documents provided to us by agency officials and stakeholders that we identified as described below. We interviewed EPA officials at headquarters and all regional offices to identify information on agency actions for managing risks. In addition, to obtain their views of EPA's actions, we interviewed former EPA officials, representatives of two associations representing state

---

[94] GAO, *Wildland Fire Management: Agencies Have Made Several Key Changes but Could Benefit from More Information about Effectiveness*, GAO-15-772 (Washington, D.C.: Sept. 16, 2015) and *Wildland Fire Risk Reduction: Multiple Factors Affect Federal-Nonfederal Collaboration, but Action Could Be Taken to Better Measure Progress*, GAO-17-357 (Washington, D.C.: May 10, 2017).

officials (the Environmental Council of States and the Association of State and Territorial Solid Waste Management Officials), a professor of environmental law, and a private consultant who has worked on Superfund issues, which we identified in the search described above and recommendations from other interviewees. We generally contacted all stakeholders that we identified who appeared to be currently working on issues related to Superfund and climate change and who agreed to speak with us. We also interviewed stakeholders at the three sites we selected as illustrative examples in order to obtain their views of EPA's actions.

We selected three nonfederal NPL sites as illustrative examples of how EPA has managed risks to human health and the environment from potential impacts of climate change effects and challenges EPA may face in managing these risks. The three sites we selected are the (1) American Cyanamid site in Bridgewater, New Jersey; (2) Iron Mountain Mine site near Redding, California; and (3) San Jacinto River Waste Pits site in Channelview, Texas. To select these sites, we initially identified 43 sites based on information in EPA documents, news articles, and interviews with EPA officials and other stakeholders as described above. We selected relevant sites in three different EPA regions that illustrate a variety of potential climate change effects and that had experienced an extreme weather event in the past 10 years. To gather more in-depth information about these sites, we reviewed EPA and other documents; toured the sites; and interviewed EPA officials and relevant stakeholders at these sites, including state and local officials, representatives of potentially responsible parties, and community organizations. The results from these illustrative examples are not generalizable to nonfederal NPL sites that we did not select.

We compared EPA's actions to manage risks to human health and the environment from the potential impacts of climate change effects with essential elements for managing risk as identified in our prior work on enterprise risk management.[95] These essential elements are as follows: (1) align the risk management process with goals and objectives, (2) identify

---

[95] GAO, *Enterprise Risk Management: Selected Agencies' Experiences Illustrate Good Practices in Managing Risk*, GAO-17-63 (Washington, D.C.: Dec. 1, 2016).

risks, (3) assess risks, (4) respond to the risks, (5) monitor the risks, and (6) communicate and report on the risks. We assessed information on EPA's actions to determine the extent to which the agency's actions aligned with these elements. In assessing EPA's actions against these essential elements, we used "aligned," "partially aligned," or "not aligned" to reflect the extent to which EPA took actions aligned with each essential element. If EPA provided evidence that it had taken major actions in alignment with that essential element, we determined the actions were aligned. If EPA provided evidence that it had taken some actions in alignment with that essential element, we determined the actions were partially aligned. If EPA took only a few or no actions in alignment with that essential element, we determined the actions were not aligned. Two GAO analysts independently reviewed the information on EPA's actions and then reached consensus on the extent to which EPA's actions were aligned with each element.

To identify the challenges EPA faces in managing these risks, we reviewed EPA documents; our prior work; and relevant documents from other organizations, including the National Research Council, that we obtained as described above. We interviewed EPA officials at headquarters and all regional offices and stakeholders in order to obtain their views on the challenges EPA faces. The views of stakeholders we interviewed are illustrative and not generalizable to all stakeholders. We reviewed the challenges that we identified in these documents and interviews and grouped all the challenges into three categories for reporting purposes: institutional, resource, and technical. Two GAO analysts independently reviewed the information and then reached consensus on the challenges and their grouping in the three categories.

We conducted this performance audit from April 2018 to October 2019 in accordance with generally accepted government auditing standards. Those standards require that we plan and perform the audit to obtain sufficient, appropriate evidence to provide a reasonable basis for our findings and conclusions based on our audit objectives. We believe that the evidence obtained provides a reasonable basis for our findings and conclusions based on our audit objectives.

APPENDIX II: COMMENTS FROM THE ENVIRONMENTAL PROTECTION AGENCY

UNITED STATES ENVIRONMENTAL PROTECTION AGENCY
WASHINGTON, D.C. 20460

OCT - 1 2019

OFFICE OF
LAND AND EMERGENCY
MANAGEMENT

Mr. Alfredo Gomez
Director
Natural Resources and Environment
U.S. Government Accountability Office
Washington, DC 20548

Dear Mr. Gomez:

Thank you for the opportunity to review and comment on the U.S. Government Accountability Office's (GAO) draft report, *EPA Should Take Additional Actions to Manage Risks from Climate Change (GAO-19-632)*. This letter provides the U.S. Environmental Protection Agency's (EPA) response to GAO's draft report findings, conclusions, and recommendations and represents coordination with other offices within the Agency.

The EPA recognizes the importance of ensuring Superfund site cleanups are resilient in the face of existing risks and extreme weather events and the Agency has taken measures to include vulnerability analyses and adaptation planning into Superfund activities. The EPA strongly believes the Superfund program's existing processes and resources adequately ensure that risks and any effects of severe weather events, are woven into risk assessments at nonfederal NPL sites. The EPA is also vigilant on a site-specific basis before and after a severe weather event occurs. Prior to a severe weather event or other upcoming risk, the EPA has a history of working with state, tribal, and local officials to identify Superfund sites that may be in the projected path. After the weather event or risk has passed, the EPA personnel are in the field to assess whether the event undermined a site's physical integrity. These site-specific steps, in addition to the Superfund program's existing processes and resources, ensure that Superfund sites are resilient in the face of existing and potential increased risks and extreme weather events.

The EPA's 2017 and 2018 experience with extreme weather demonstrated that the Agency's efforts are working to address extreme weather challenges. After 2017's very active hurricane season, the EPA performed an analysis of Superfund NPL and Superfund Alternative Approach site remedies. This comprehensive analysis found overall that remedies are resilient. While tropical force winds or greater and/or inundation affected more than 250 sites, only 16 sites reported minor damage, and none indicated impairment to a remedy's protectiveness. Similarly, after Hurricane Florence in 2018, the EPA coordinated post-landfall assessments of approximately 50 Superfund sites in North Carolina and found no need for post-storm action.

The Agency appreciates GAO's review and updates to EPA's 2012 analysis evaluating NPL site remedy vulnerabilities (the 2012 report was the basis for the Superfund Climate Resilience Action Plan, which is in the implementation phase). Both the 2012 EPA report and the 2019 GAO report recognize that a significant percentage (60 percent) of nonfederal NPL sites are in locations that may be affected by risks and severe weather events, that may increase in intensity, duration, or frequency. However, again, the EPA strongly believes the Superfund program's existing processes and resources adequately ensure that risks and any effects of severe weather events, that may increase in intensity, duration, or frequency, are woven into risk assessments at nonfederal NPL sites.

Below are the EPA's comments on the GAO recommendations.

GAO Recommendation 1
The Director of OSRTI should establish a schedule for completing the standardization and improvement of information on the boundaries of nonfederal NPL sites.

EPA Response
The EPA agrees with this finding and has already taken the initial steps prior to this GAO review to develop an approach to improve and standardize Superfund nonfederal NPL site boundaries. The approach will be a substantial and resource-intensive effort requiring input from every NPL or Superfund Alternative Approach site remedial project manager and potentially may require current and former contractors' input. The Agency anticipates completing the analysis to set milestones by fiscal year 2020, Quarter 2 with the aim to have collected an initial set of site boundaries for all NPL sites by fiscal year 2021, Quarter 4. While the EPA acknowledges the lack of consistent standards to identify site boundaries exists on the national level, current efforts on a regional and site-specific basis take into account measures that allows the Agency to assess the risks and identify site boundaries. Additionally, the method of determining site boundaries is only one aspect that can affect the Agency's capacity to address site risks, and the EPA strongly believes the Superfund program's existing processes and resources adequately ensure that risks and any effects of severe weather events, that may increase in intensity, duration, or frequency, are woven into risk assessments at nonfederal NPL sites. These site-specific steps, in addition to the Superfund program's existing processes and resources, ensure that Superfund sites are resilient in the face of existing and potential increased risks and extreme weather events.

GAO Recommendation 2
The EPA Administrator should clarify how Agency actions to manage risks to human health and the environment from the potential impacts of climate change effects at non-federal NPL sites align with the Agency's current goals and objectives.

EPA Response
The EPA understands this recommendation's intent, but the Agency has previously acknowledged its commitment to aligning its risk process with the mission of the Agency and its goals and objectives. The EPA believes that managing risks from exposure to contaminants in the environment is integral to the EPA's current Strategic Goal 1.3, Revitalize Land and Prevent Contamination. The Agency has taken measures to include vulnerability analyses and adaptation planning into the Superfund program and activities, taking the steps to ensure Superfund sites are

resilient and allowing the EPA to effectively manage risks to human health and the environment, consistent with the Agency's mission. Additionally, as we have previously detailed to GAO, the EPA is currently implementing an approach by the Agency to build and strengthen the adaptive capacity of its partners in states, tribes, and local communities through multiple facets—training to increase awareness of the risks posed by climate change to EPA's programs (including OLEM and the Superfund program); tools to provide necessary data, information, and tools, including tools to empower clean-up mangers to assess the risks posed by climate change to Superfund sites; and financial incentives to support climate-resilient investments in communities across the nation. The EPA believes existing processes, resources, and commitments adequately ensure that the Agency's risk management process aligns with current goals and objectives—fulfilling the GAO's identified essential element of enterprise risk management—while also providing a basis for the Agency's view that the Superfund program's existing processes and resources adequately ensure that risks and any effects of severe weather events, that may increase in intensity, duration, or frequency, are woven into risk assessments at nonfederal NPL sites. These additional measures, in addition to the Superfund program's existing processes and resources, ensure that Superfund sites are resilient in the face of existing and potential increased risks and extreme weather events.

The EPA strongly believes that we have reasonable assurance throughout the Agency that the current ongoing actions to manage these risks at nonfederal NPL sites align with current Agency goals and objectives, including throughout senior officials and leadership who take an active role in supporting these actions and help the EPA achieve its mission of protecting human health and the environment.

<u>GAO Recommendation 3</u>
The Director of OSRTI should provide direction on how to integrate information on the potential impacts of climate change effects into risk assessments at nonfederal NPL sites.

<u>EPA Response</u>
The EPA strongly believes the Superfund program's existing processes and resources adequately ensure that risks and any effects of severe weather events, that may increase in intensity, duration, or frequency, are woven into risk assessments at nonfederal NPL sites. While the EPA believes that current Agency efforts fulfill the basis for the recommendation, to buttress awareness, the OSRTI office director may issue a memorandum to reinforce for site teams the array of tools and resources available to them. A determination to this effect will be made by January 30, 2020. This additional reinforcement would supplement the Superfund program's existing processes and resources that already ensure that Superfund sites are resilient in the face of existing and potential increased risks and extreme weather events.

<u>GAO Recommendation 4</u>
The Director of OSRTI should provide direction on how to integrate information on the potential impacts of climate change effects into risk response decisions at nonfederal NPL sites.

<u>EPA Response</u>
The EPA strongly believes the Superfund program's existing processes and resources adequately ensure that risks and any effects of severe weather events, that may increase in intensity,

duration, or frequency, are woven into risk response decisions at nonfederal NPL sites. The OSRTI office director may reiterate existing authorities and tools for considering extreme weather effects through the memorandum described in the response to Recommendation 3. This additional reinforcement would supplement the Superfund program's existing processes and resources that already ensure that Superfund sites are resilient in the face of existing and potential increased risks and extreme weather events.

Thank you for the opportunity to review the draft report. We believe the report contains information that will strengthen the EPA's continued efforts to build resilience into Superfund sites. In addition to our comments found in this letter, we have also enclosed technical comments on the draft report. If you have any questions or need additional information, for the Office of Land and Emergency Management, please contact Kecia Thornton at (202) 566-1913 and for the Office of the Administrator, Michael Benton at (202) 564-2860.

Sincerely,

Peter C. Wright
Assistant Administrator

Enclosure

cc: Barry Breen, OLEM
Steven Cook, OLEM
James Woolford, OLEM
Dana Stalcup, OLEM
Kent Benjamin, OLEM
Travis Voyles, AO
Kevin DeBell, AO
Lance McCluney, AO
Michael Benton, AO
EPA GAO Liaison Team

In: Climate Change
Editor: Iyana Hendricks

ISBN: 978-1-53617-757-2

Chapter 6

DRILLING UNDER NEW WYOMING LEASES ON HOLD AS JUDGE REQUIRES BLM TO RECONSIDER IMPACTS ON CLIMATE CHANGE*

Adam Vann

A recent decision in the United States District Court for the District of Columbia Circuit has paused oil and gas exploration and production activity in certain leased areas of Wyoming and hinted at heightened requirements that the Bureau of Land Management (BLM) must satisfy to comply with the National Environmental Policy Act (NEPA) before issuing oil and gas leases. Specifically, the decision will require BLM to conduct a more thorough review of the potential climate change impacts of certain oil and gas leases before allowing the lessees to conduct drilling operations.

---

* This is an edited, reformatted and augmented version of Congressional Research Service Publication No. LSB10289, dated April 16, 2019.

The decision addresses five oil and gas lease sales for areas in Wyoming between May 2015 and August 2016 that resulted in BLM issuing 282 leases. BLM had determined that the lease sales did not require it to issue an Environmental Impact Statement (EIS), which is a detailed evaluation of a federal action that NEPA and its implementing regulations require for any federal action that will "significantly affect the quality of the human environment." Instead, BLM issued Environmental Assessments (EAs) that were "tiered to," i.e., incorporated, a resource management plan for the relevant area and an EIS, which BLM had produced at the land use planning stage. Addressing how the leases would impact climate change on a "conceptual level," the EAs acknowledged that oil and gas drilling on the parcels to be leased would cause Greenhouse Gas (GHG) emissions. However, BLM did not attempt to quantify and project the GHG emissions likely to result from the lease sales. The EAs noted that "[t]he inconsistency in results of scientific models … limits the ability to quantify potential future impacts at this level." The Eas concluded that the GHG emissions likely to result from each lease sale "would not have a measurable effect" when compared to national or global emissions, or would represent only an "incremental contribution" to regional and global GHG emissions. For each of the EAs, BLM issued a Finding of No Significant Impact (FONSI), meaning that BLM did not require an EIS to issue the leases.

Non-profit groups, including WildEarth Guardians, challenged BLM's issuance of these leases, arguing that "NEPA required BLM to conduct a more piercing consideration of the consequences of oil and gas drilling before it authorized the Wyoming Lease Sales." More specifically, Plaintiffs argue[d] that, by issuing the EAs and FONSIs in conjunction with the Wyoming Lease Sales, BLM violated NEPA because it failed to take a "hard look" at the GHG emissions and their impact on climate change. The court addressed a number of specific alleged shortcomings in the agency's efforts to consider the direct, indirect and cumulative impacts of the lease sales on GHG emissions as required by NEPA and its implementing regulations.

The court noted that once the leases were sold, BLM had made an "irrevocable commitment" to allow at least some GHG emissions, and therefore had to analyze the reasonably foreseeable impacts of those leases on GHG emissions. The court acknowledged that BLM was not required to assess or evaluate impacts on a parcel-by-parcel basis at the lease sale stage, and that a simple qualitative analysis of GHG emissions impacts in an EA is acceptable if more definitive information cannot be provided. However, the court found that "while BLM could not, at the leasing stage, reasonably foresee the environmental impacts of specific drilling projects, it could reasonably foresee and forecast the impacts of oil and gas drilling across the leased parcels as a whole." The court thus concluded that "given the mix of information available to BLM at the leasing stage, NEPA required that BLM reasonably quantify the GHG emissions resulting from oil and gas development on the leased parcels in the aggregate."

The court also found that BLM had "failed to sufficiently consider another potential effect of leasing: the downstream GHG emissions that would result from the production of oil and gas on the leased parcels." As the court noted, NEPA regulations require agencies to evaluate the indirect effects that are "caused by [a proposed federal] action and are later in time or farther removed in distance, but are still reasonably foreseeable," and the scope of this requirement has been a subject of considerable debate in recent years. After a lengthy review of existing precedent, the court found that "even under the heightened causation standard established" in previous cases, "downstream GHG emissions from fossil fuel use are an indirect effect of BLM's oil and gas leasing program at issue here" and therefore BLM was required to consider them in its NEPA review of the lease sales.

The court next considered the plaintiff's claim that BLM had not adequately addressed the *cumulative* impacts of the lease sales on GHG emissions as required by NEPA regulations. The court agreed with the plaintiffs, specifically noting that BLM's failure to quantify estimated GHG emissions undermined its conclusion that emissions resulting from the leases merely "represent an incremental contribution to the total regional and global GHG emissions levels." The court held that without quantified estimates, BLM "could not conceptualize the extent to which the lease sales would contribute to the local, regional and global climate change discussed in the EAs and tiered EISs." However the court disagreed with the plaintiffs' assertion that BLM should be required to employ a methodology that monetizes GHG emissions in the context of a "global carbon budget" in order to quantify the "social cost of carbon," finding that the agency's decision to forgo this analysis "does not rise to the level of a NEPA violation."

While this decision is important because of the court's determinations regarding the extent of agency obligations under NEPA to consider GHG emissions prior to lease sales, the court's choice of remedy is also noteworthy. The plaintiffs asked the court to void the Wyoming leases and enjoin BLM from authorizing leases until the agency complied with NEPA requirements per GHG emissions. However, while the court sided with the plaintiffs on the issue of BLM's NEPA obligations, it chose not to void the leases. Instead the court remanded the non-conforming EAs and FONSIs to BLM to allow the agency to address the deficiencies identified in the decision. The court noted that these deficiencies consist "merely of a failure to fully discuss the environmental effects of these lease sales" and that "nothing in the record indicates that on remand the agency will necessarily fail to justify its decisions to issue EAs and FONSIs." Noting, however, that the failures raised doubts about BLM's decision making regarding the Wyoming leases, the court enjoined BLM from issuing any permits to drill under the leases at issue in the case until BLM addresses its NEPA compliance shortcomings.

It is not yet known if BLM will successfully appeal this decision. If the decision stands, it has policy implications that may be of interest to Congress. By holding that NEPA requires BLM to quantify how its proposed leases impact GHG emissions, the decision hints that NEPA may require federal agencies to consider the impacts of a broader scope of federal actions on GHG emissions. Congress could amend NEPA to add specificity to its requirements, or look into NEPA compliance at BLM and other federal agencies.

In: Climate Change
Editor: Iyana Hendricks
ISBN: 978-1-53617-757-2

Chapter 7

THE UNITED NATIONS FRAMEWORK CONVENTION ON CLIMATE CHANGE, THE KYOTO PROTOCOL, AND THE PARIS AGREEMENT: A SUMMARY*

Jane A. Leggett

SUMMARY

The United Nations Framework Convention on Climate Change (UNFCCC) has been the principle forum for cooperation among nations on greenhouse gas (GHG)-induced climate change since its adoption in 1992. Its objective is "to stabilize greenhouse gas concentrations in the atmosphere at a level that will prevent dangerous human interference with the climate system, in a time frame which allows ecosystems to adapt naturally and enables sustainable development."

---

* This is an edited, reformatted and augmented version of Congressional Research Service Publication No. R46204, dated January 29, 2020.

Stabilizing GHG concentrations in the atmosphere requires that the balance of "gross" emissions of GHG minus the removals of GHG from the atmosphere reach "net zero."

Two principles agreed in the UNFCCC are that (1) Parties should act "on the basis of equity and in accordance with their common but differentiated responsibilities and respective capabilities" and (2) developed country Parties should take the lead in combating climate change. The bifurcation of responsibilities among Parties into developed (Annex I) and developing countries has been a major point of contention. Annex I Parties, including the United States, had stronger obligations, such as more rigorous reporting and reviews. A subset listed in Annex II, including the United States, committed to provide agreed financial resources and technology transfers. The commitments are qualitative and collective, not binding on individual Parties.

The first subsidiary agreement to the UNFCCC was the 1997 Kyoto Protocol (KP), which entered into force in 2005. The United States signed but did not ratify the KP and so is not a Party. The developed Parties agreed to reduce GHG emissions by 5% below their 1990 levels, with different targets for each Party.

In 2009, a political declaration, the Copenhagen Accord, led to explicit pledges from many Parties to mitigate GHG, though they remained bifurcated as Annex I and non-Annex I (i.e., developing countries) by both the type of action and the frequency and format of the reporting requirements. In 2010, the Cancun agreements took note of a Copenhagen pledge by developed country Parties to jointly mobilize $100 billion per year by 2020. Funds provided "may come from a wide variety of sources, public and private, bilateral and multilateral, including alternative sources."

The Paris Agreement (PA) is the second major subsidiary agreement under the UNFCCC. The PA defines a collective, long- term objective to hold the GHG-induced increase in temperature to well below 2o Celsius (C) and to pursue efforts to limit the temperature increase to 1.5o C above the pre-industrial level. In the PA, for the first time under the UNFCCC, all

Parties participate in a common framework with common guidance, though some Parties are allowed limited flexibility.

The negotiators intended the PA to be legally binding on its Parties, though not all provisions are mandatory. All Parties must submit "Nationally Determined Contributions" (NDCs) containing nonbinding pledges to mitigate GHG emissions. The Parties are to update or submit new NDCs by 2020 and every five years thereafter. Each successive NDC of a Party "will represent a progression" and "reflect its highest possible ambition, reflecting its common but differentiated responsibilities and respective capabilities, in light of different national circumstances."

The PA reiterates the obligation in the UNFCCC for developed country Parties to seek to mobilize financial support to assist developing country Parties with climate change mitigation and adaptation efforts, encouraging all Parties to provide financial support voluntarily. The decision to carry out the PA calls for continuing the Cancun collective mobilization through 2025. The Parties agree to set, prior to their 2025 meeting, a new collective, quantified goal of not less than $100 billion annually to assist developing country Parties.

President Trump announced his intention in 2017 to withdraw the United States from the PA as soon as it was eligible. The U.S. Department of State notified the United Nations of U.S. withdrawal on November 4, 2019. The withdrawal takes effect on November 4, 2020, unless the U.S. government postpones or rescinds the withdrawal. A Party may reenter the PA 30 days after depositing notice that it has ratified, accepted, or acceded to the PA.

INTRODUCTION

Multiple decades of scientific studies find that human activities induce global climate change by emitting greenhouse gases (GHGs) from fuel combustion, certain industries, deforestation, and other activities. Scientists researched and assessed the science of GHG-induced climate change for more than 150 years before government policymakers around the world agreed to cooperate to consider how to address its risks to humans and ecosystems.[1] Following several international scientific meetings in 1985-1987, governments decided to establish the Intergovernmental Panel on Climate Change (IPCC), under the auspices of the United Nations Environment Programme and the World Meteorological Organization, to provide them with assessments of climate change science, projected social and economic impacts, and potential response strategies.[2] In 1989, the U.N. General Assembly provided a mandate to negotiate what became, in 1992, the U.N. Framework Convention on Climate Change (UNFCCC).[3]

The UNFCCC has been the primary multilateral vehicle since 1992 for international cooperation among national governments to address GHG-induced climate change. While the UNFCCC is a focal point for national governments, its periphery is one forum, among others, for information sharing, collaboration, and activism also for subnational governments, financial institutions, the private sector, and nongovernmental organizations. This chapter is not describing these other, increasingly important aspects of international cooperation on climate change.

This chapter summaries the content of the UNFCCC and its two subsidiary international treaties: the 1997 Kyoto Protocol (KP) and the 2015 Paris Agreement (PA). It also describes the existing guidelines to implement the PA, known as the 2018 Katowice Climate Package. The report highlights information relevant to the 2019 climate change conference, known as COP25. This chapter is not comprehensive. A number of other CRS reports provide greater detail and nuance on these and other aspects of the international climate change negotiations and cooperation. Some are listed at the end of this chapter.

---

[1] See CRS Report R45086, Evolving Assessments of Human and Natural Contributions to Climate Change, by Jane A. Leggett.

[2] Notable scientific conferences included the International Conference on the Assessment of the Role of Carbon Dioxide and of Other Greenhouse Gases in Climate Variations and Associated Impacts at Villach, Austria, in October 1985, as well as workshops in Villach, Austria, and Bellagio, Italy, in 1987. See the IPCC's history at https://www.ipcc.ch/about/history/.

[3] U.N. General Assembly, "Protection of Global Climate for Present and Future Generations," Resolution 43/53, 1989.

THE U.N. FRAMEWORK CONVENTION ON CLIMATE CHANGE

The UNFCCC has been the primary multilateral vehicle since 1992 for international cooperation to address GHG-induced climate change.[4] As of January 1, 2020, there are 197 Parties to the UNFCCC that have ratified, accepted, or acceded to the international treaty, including the United States.[5] There is broad agreement that participation of all countries would be necessary to achieve the objective of the UNFCCC, which is stated as follows:

> to stabilize greenhouse gas concentrations in the atmosphere at a level that will prevent dangerous human interference with the climate system, in a time frame which allows ecosystems to adapt naturally and enables sustainable development.[6]

Achieving the objective would require both abatement of GHG emissions and facilitation of adaptation to adverse impacts of climate change in order to enable sustainable development. Stabilizing GHG concentrations in the atmosphere requires that net GHG emissions—the balance of "gross" emissions of GHG to the atmosphere and removals of GHG from the atmosphere— reach "net zero" or "carbon neutrality." Removals and sequestration can occur by photosynthesis (vegetation, sea algae) or through advanced technologies.[7] Some increased level of removals, or "sinks," could allow for some amount of human-related GHG

---

[4] UNFCCC, agreed on May 9, 1992, entered into force March 21, 1994, 1771 U.N.T.S. 107; United Nations, *Treaty Series*, vol. 1771, p. 107; and U.S. depositary notifications C.N.148.1993.

[5] U.N. Treaty Collection, Chapter XXIII. 7. President George H. W. Bush transmitted the signed treaty to the Senate for its advice and consent in 138 *Congressional Record* 23902 (September 8, 1992). The U.S. Senate gave its advice and consent to ratification in 138 *Congressional Record* 33527 (October 7, 1992). See also S. Treaty Doc. 102-38 (1992); S. Exec. Rept. 102-55. President Bush signed the instrument of ratification and submitted it to the United Nations on October 13, 1992. Depositary notification C.N.148.1993.

[6] UNFCCC, Article 2.

[7] See, for example, CRS Report R41371, *Geoengineering: Governance and Technology Policy*, by Kelsi Bracmort and Richard K. Lattanzio.

emissions to continue. The United States and other Parties to the UNFCCC agreed to this objective when they ratified the treaty.

As a framework convention, this international treaty provides the structure for collaboration and evolution of efforts over decades, as well as the first qualitative step in that collaboration. The UNFCCC does not, however, include quantitative and enforceable objectives and commitments for any Party.

The UNFCCC was adopted in 1992 and entered into force in 1994. The UNFCCC's governing body, the Conference of the Parties (COP), met in its 25th session (COP25) from December 2 to 13, 2019, in Madrid, Spain. Initially, Chilean President Sebastián Piñera stepped forward to host COP25 in place of Brazil following the election of President Jair Bolsonaro.[8] Piñera sought to underscore his efforts to address climate change but ultimately decided that the summit should take place elsewhere due to mass protests in Chile.[9]

All Parties to the UNFCCC, including the United States, have a set of common obligations under the treaty:

- to inventory, report, and mitigate their human-related GHG emissions, including emissions and removals from land uses;
- to cooperate in preparing to adapt to climate change; and
- to assess and review, through the COP, the effective implementation of the UNFCCC, including the commitments therein.

Certain obligations are additional or more specific for the countries that had higher incomes in 1992, and those countries are listed in Annex I of the Agreement. They are commonly referred to as Annex I Parties. All others are non-Annex I Parties. These additional or more specific

---

[8] Reportedly, Brazil's foreign minister notified the UNFCCC in November 2018 that Brazil would not host the conference. UNFCCC, "Statement on COP25 Host," November 30, 2018, https://unfccc.int/news/statement-on-cop25- host.

[9] See, for example, Ernesto Londoño and Somini Sengupta, "Chile, Rocked by Unrest, Withdraws from Hosting Climate and Trade Summits," *New York Times*, November 1, 2019, https://www.nytimes.com/2019/10/30/world/ americas/chile-cop25-apec.html.

obligations included more frequent reporting and providing financing and technology transfers, among others.

The bifurcation of Parties and commitments has been a major point of contention and, some would argue, delay in negotiation and implementation of the climate change agreements (see text box). The UNFCCC and its subsidiary agreements do not define the terms developing country or developed country. In the 1990s, the Annex I Parties anticipated that developing country Parties would "graduate" into specific commitments and become donor countries as their incomes and emissions grew. As discussed later, related disagreements directly contributed to U.S. nonparticipation in the KP, the collapse of negotiations in Copenhagen in 2009, and the withdrawal or decision of some Parties not to adopt GHG abatement targets in the second period of the KP from 2013 to 2020.

The Copenhagen Accord

In Copenhagen at COP15 in 2009, the COP was unable to adopt an agreement among all Parties as Bolivia, Cuba, Peru, and Venezuela opposed the text. The decision of the COP included a nonbinding political statement, the Copenhagen Accord,[10] which began a turn toward more explicit commitments by non-Annex I Parties to GHG mitigation under the UNFCCC. The Copenhagen Accord specified that the Annex I Parties would implement quantified economy- wide GHG targets for 2020 in an agreed reporting format. Non-Annex I Parties to the UNFCCC would commit to implement mitigation actions to be submitted in an alternative agreed format.

---

[10] "The Copenhagen Accord," in Report of the Conference of the Parties on Its Fifteenth Session, Held in Copenhagen from 7 to 19 December 2009 - Addendum: Part Two: Action Taken by the Conference of the Parties at Its Fifteenth Session: Decisions Adopted by the Conference of the Parties, Decision 1/CP.15, UNFCCC, 2010, https://unfccc.int/process-and-meetings/conferences/past-conferences/copenhagen-climate-change-conference-december-2009/cop-15/ cop-15-decisions.

At least 43 Annex I Parties (15 Parties, including the United States, plus the EU-28 jointly submitting a pledge) and 47 non-Annex I Parties had submitted nonbinding pledges.[11] While most countries participated, the pledges remained bifurcated by both the type of action and the reporting requirements. Among other differences, Annex I Parties were to submit quantified economy-wide GHG emissions targets for 2020 relative to a baseyear, while non-Annex I Parties were to submit "nationally appropriate mitigation actions" with no associated dates. The submissions would be compiled separately by the Secretariat of the UNFCCC.

Continuing Challenge: Common but Differentiated Responsibilities

Two principles in the UNFCCC are that (1) Parties should act "on the basis of equity and in accordance with their common but differentiated responsibilities and respective capabilities" and (2) that developed country Parties should take the lead in combating climate change.

Thus, the 1992 UNFCCC bifurcated Parties into two broad categories: (1) the 35 highest-income countries at that time, which were listed in Annex I and accepted more specific commitments regarding national plans, reporting, and assisting low-income countries; and (2) all other Parties—the non-Annex I Parties.

As a subset, Annex II Parties are those Annex I Parties (primarily members of the Organisation for Economic Co- operation and Development at that time) that took on further responsibilities (Article 4.3 and 4.4) to provide new and additional financial resources to meet the agreed full costs incurred by "developing country Parties" in complying with the obligation for reporting their emissions, policies, and measures (as listed in Article 12). Annex II Parties also agreed to provide "financial resources, including for the transfer of technology, to meet the agreed full incremental costs in implementing measures" to meet general, common, and unenforceable commitments (as listed in Article 4.1).

The UNFCCC recognizes that the extent to which developing country Parties implement their commitments under the UNFCCC would depend on

[11] Katarina Buhr et al., *Comparisons of the Copenhagen Pledges*, Centre for Climate Science and Policy Research, 2012, http://www.diva-portal.org/smash/record.jsf?pid=diva2%3A570663&dswid=1644.

effective implementation by "developed country Parties" of the commitments regarding financial resources and transfer of technology. The term non-Annex I Parties is often treated synonymously with developing country Parties—though a number of non-Annex I Parties are now, more than 25 years later, classified by the World Bank as middle- to high-income.

Because GHG emissions come from sources in all countries, only limitations then reductions by all major emitters can stabilize the rising GHG concentrations in the atmosphere. The UNFCCC states that the developed country Parties would take the lead in abating GHG emissions. Many of the delegations negotiating the treaty believed that the non-Annex I Parties would follow with significant GHG mitigation thereafter. By 1995, however, a position of "no new commitments for developing countries" from some leading non-Annex I Parties effectively blocked non-Annex I Parties' participation in quantitative mitigation commitments—even by willing emerging economies—in the KP and prevented a legal agreement in Copenhagen in 2009.

A turning point occurred in the 2009 Copenhagen COP, when the non-binding Copenhagen Accord invited GHG mitigation pledges, albeit differently described, from all Parties. Subsequently, the submission before COP21 in 2015 by China and many other non-Annex I Parties of intended nationally determined contributions (NDCs) contributed to the successful negotiation of the PA in 2015.

Arguably the most significant outcome of the PA is that, for the first time under the UNFCCC, all Parties agreed to participate in a common framework with common guidance, though some Parties are allowed flexibility in line with their capacities. This largely supersedes the bifurcated mitigation obligations of developed and developing countries that held the negotiations in often-adversarial stasis for many years. Still, many discussions under the PA return to debates about how to take into account equity and the principle of common but differentiated responsibilities and respective capabilities in light of different national circumstances.

The Kyoto Protocol (KP)

The first subsidiary agreement to the UNFCCC was the 1997 KP, which entered into force in 2005.[12] The United States signed but did not ratify the KP and so is not a Party to it. The KP established legally binding targets for 37 high-income countries and the European Union (EU) to reduce their GHG emissions on average by 5% below 1990 levels during 2009-2012. It precluded GHG mitigation obligations for developing countries.

All Parties with the Quantified Emissions Limitation and Reduction Obligations (QELROS) under the KP (i.e., GHG targets) were judged in compliance after the end of the first commitment period of 2009-2012.[13] The domestic GHG emissions of some Parties were higher than their targets, but as envisioned under the KP, Parties could fulfil their obligations by acquiring emission reduction credits through the three market mechanisms of the treaty: the Clean Development Mechanism, Joint Implementation, and emissions trading.[14]

Most of the high-income Parties—mostly the EU members and other European nations—took on further GHG reduction targets for 2013-2020. The Secretariat's assessment of the emissions of the KP Parties with QELROS, as of November 2018, found:

---

[12] U.N. Treaty Collection, Chapter XXVII Environment, 7.1, "Kyoto Protocol to the United Nations Framework Convention on Climate Change," December 11, 1997, https://treaties.un.org/Pages/ViewDetails.aspx?src=TREATY&mtdsg_no=XXVII-7-a&chapter=27&clang=_en.

[13] UNFCCC, "Assessment of First Phase of Kyoto Protocol," https://unfccc.int/news/assessment-of-first-phase-of- kyoto-protocol-published. Several studies found that the Clean Development Mechanism (CDM) approvals of GHG mitigation credits were likely corrupt due to "inappropriate validation of projects, including overestimation, double- counting, or fraudulent trade of carbon credits." Transparency International, "Global Corruption Report: Climate Change—A User's Guide," 2011. A study commissioned by the EU confirmed problems, leading to a decision by the EU that it would not accept CDM credits to count against member state targets, leading to a collapse of credit prices. Öko-Institut e.V., "How Additional Is the Clean Development Mechanism? Analysis of the Application of Current Tools and Proposed Alternatives," March 2016, https://ec.europa.eu/clima/sites/clima/files/ets/docs/ clean_dev_mechanism_en.pdf. This is leading many Parties to negotiate for stricter rules for validating and reporting use of emissions trading under the PA, as discussed later in this chapter.

[14] UNFCCC, "Mechanisms Under the Kyoto Protocol," https://unfccc.int/process/the-kyoto-protocol/mechanisms.

> Annex I Parties are progressing towards their 2020 targets but gaps remain. Individual Parties have made varying progress towards their 2020 targets: most Parties' emission levels are already below their 2020 targets; some Parties must make further efforts to meet their targets by strengthening implementation of their existing [policies and measures]; and using units from MBMs [market-based mechanisms], if needed, and the contribution from LULUCF [land use, land use change, and forestry], if applicable; other Parties' emissions remained above their base-year level, owing mainly to inadequacy of domestic [policies and measures], high marginal mitigation costs or energy system constraints—they indicated that the use of units from MBMs and, if applicable, the contribution from LULUCF are expected to make a sizable contribution towards achieving their targets.[15]

The United States did not join the KP, and Canada withdrew before the end of the first commitment period. At least in part, their reasons for disengaging from the KP included the non- Annex I Parties' objections to acceding to quantified GHG reduction commitments. While negotiating the second KP commitment period, Australia, Japan, and other Parties also decided to seek an agreement that included commitments on the same terms from all Parties. This led to a mandate, negotiated at the 2011 COP17 in Durban, South Africa, to develop a protocol, another legal instrument, or an agreed outcome with legal force under the UNFCCC applicable to all Parties no later than 2015. The Durban Mandate resulted in the 2015 PA, discussed below.

The Paris Agreement (PA)

The PA is the second major subsidiary agreement under the UNFCCC. The PA is to eventually replace the KP as the primary subsidiary vehicle for process and actions under the UNFCCC. Obama Administration

---

[15] Subsidiary Body for Implementation, "Compilation and Synthesis of Third Biennial Reports of Parties Included in Annex I to the Convention: Executive Summary," FCCC/SBI/2018/INF.8, November 22, 2018.

officials stated that the PA is not a treaty requiring Senate advice and consent to ratification.[16]

The U.N. Climate Conference in Madrid included COP25 and the second session of the "Conference of the Parties serving as the meeting of the Parties to the Paris Agreement" (CMA2), along with meetings of other related bodies. Though the United States has given notice of withdrawal from the PA, its withdrawal is to take effect no earlier than November 4, 2020. Until then, the United States may participate as a Party. After withdrawal takes effect, the United States may participate in a more limited way as an Observer State.

The PA was intended to be legally binding on its Parties, though not all provisions in it are mandatory. The PA requires that Parties submit nonbinding pledges, in NDCs, to mitigate their GHG emissions and enhance removals. NDCs may also articulate goals to adapt to climate change and cooperate toward these ends, including mobilization of financial and other support. Some provisions are binding, such as those regarding reporting and review, while others are recommendations or collective commitments to which it would be difficult to hold an individual Party accountable. Key aspects of the agreement include:

- *Temperature goal.* The PA defines a collective, long-term objective to hold the GHG-induced increase in temperature to well below 2° Celsius (C) and to pursue efforts to limit the temperature increase to 1.5^{o} C above the pre-industrial level. As discussed below, a periodic "Global Stocktake" is to assess progress toward the goals.
- *Single GHG mitigation framework.* The PA establishes a process, with a ratchet mechanism in five-year increments, for all countries to set and achieve GHG emission mitigation pledges until the long-term goal is met. For the first time under the UNFCCC, all Parties

[16] U.S. Department of State, "Background Briefing on the Paris Climate Agreement," December 12, 2015, https://2009- 2017.state.gov/r/pa/prs/ps/2015/12/250592.htm. For example, "In terms of congressional approval, this agreement does not require submission to the Senate because of the way it is structured. The targets are not binding; the elements that are binding are consistent with already approved previous agreements."

participate in a common framework with common guidance, though some Parties are allowed flexibility in line with their capacities.

- *Accountability framework.* To promote compliance, the PA balances accountability to build and maintain trust (if not certainty) with the potential for public and international pressure ("name-and-shame"). Also, the PA establishes a compliance mechanism designed to use expert-based and facilitative review and response rather than punitive measures. Many Parties and observers are to closely monitor the effectiveness of this strategy.
- *Adaptation.* The PA also requires "as appropriate" that Parties prepare and communicate their plans to adapt to climate change. Parties agreed that adaptation communications would be recorded in a public registry.
- *Collective financial obligation.* The PA reiterates the collective obligation in the UNFCCC for developed country Parties to provide financial resources—public and private—to assist developing country Parties with mitigation and adaptation efforts. It urges scaling up from past financing. The Parties agreed to set, prior to their 2025 meeting, a new collective quantified goal for mobilizing financial resources of not less than $100 billion annually to assist developing country Parties.

The Katowice Package

At COP24/CMA1 in Katowice, Poland, in 2018, the PA Parties agreed to many of the guidelines and processes so that Parties may implement the PA as intended. Despite these agreements, Parties did not resolve several issues of significance. Negotiations on these issues will likely continue at COP26/CMA3 in Glasgow, Scotland, in November 2020. The Katowice Package, as it is often called, clarified some ambiguities in the PA that were considered important to U.S. interests, including guidelines for Parties to report their NDCs, and the Enhanced Transparency Framework

(ETF) with guidelines and formats to allow a Party's NDC to be clearly understood. The Katowice Package thereby supports the effectiveness of the consultative compliance mechanism of the PA (discussed below). Below are brief summaries of key aspects of the Katowice Package.

NDC Guidelines

The Parties to the PA agreed to new guidelines on how to report NDCs. NDCs are to be updated every five years and "will" represent a progress in ambition to abate GHG emissions beyond the previous NDC. While Parties agreed that they "should" use a prescribed format for communicating NDCs, the details are still to be worked out. There is not agreement yet on "common timeframes" for NDCs—whether NDCs should look five or 10 years into the future. Those Parties that submitted NDCs with time frames up to 2025 (including the United States) must communicate "new" NDCs by 2020.[17] Those Parties with NDCs with time frames up to 2030 must communicate or update their NDC in 2020. Were the United States to remain in the PA, it would be required to submit a new NDC in 2020.

The content of NDCs continues to be nationally determined and nonbinding, but it should reflect what a Party intends to achieve. The guidelines apply to NDCs submitted in 2025, but Parties are invited to use an agreed format in updating their NDCs in 2020. The guidelines also address how to report adaptation measures for Parties that wish to include them in their NDCs.

Voluntary Cooperation and Market Mechanisms

The PA provides in Article 6 for Parties to choose voluntary cooperation with other Parties to implement their NDCs. The purpose is to allow "higher ambition in their mitigation and adaptation actions and to

[17] Parties to the UNFCCC submitted Intended Nationally Determined Contributions before or after COP21 in Paris, before the PA's entry into force on November 4, 2016. Almost all of the INDCs became NDCs without resubmission by Parties. Parties had not anticipated such rapid entry into force, and the PA was written so that Parties should submit their NDCs by 2020. Consequently, Parties must submit their NDCs by 2020, though they could be identical to existing NDCs, could be new in some aspects, and could be "enhanced"—increasing the ambition of the GHG mitigation pledge therein.

promote sustainable development and environmental integrity" (Article 6.1). This article, in other words, allows the use of market mechanisms to achieve GHG mitigation at the lowest possible cost and in concert with sustainable development. This, in theory, can induce Parties to take on stronger GHG mitigation commitments while ensuring that the GHG mitigation constitutes real emission reductions. The debate about the purpose for voluntary cooperation and market mechanisms—and the rules by which they are put into operation—was a major area of work undecided in Katowice.

Adaptation Reporting

Parties agreed to provide information on adaptation priorities, needs, plans, and actions in new "adaptation communications," as well as through the NDCs.

Parties agreed in Katowice that the Adaptation Fund, originally established under the 1997 Kyoto Protocol, will serve the PA. It will be one of the operating entities to the financial mechanism of the PA in addition to the Global Environment Facility and the Green Climate Fund, discussed below.

Global Stocktake

As prescribed by the PA, a Global Stocktake is to be held every five years. Parties agreed that the Global Stocktake will consider progress toward the UNFCCC's objective and the PA's aims overall. It will use best available science and will cover mitigation, adaptation, financial flows, equity, and means of implementation and support. It will not examine the situations of individual Parties. Parties decided that the next Global Stocktake would be held in 2023.

A number of decisions were reached regarding the Global Stocktake, including the information it is expected to receive from the ETF (discussed below) and other sources from PA processes.

Input may also come from nonstate actors, including non-Parties, localities and subnational governments, the business community, and all parts of civil society.

The Enhanced Technology Framework (ETF)

Setting strong requirements for the transparency of each Party's efforts has been a priority of the United States since the negotiation of the UNFCCC. ETF guidelines specify the information that Parties must report with their NDCs. That information is expected to support a "facilitative multilateral consideration of progress," along with biennial transparency reports. Methods for GHG emission estimation and other technical issues will continue to rely on the IPCC's technical advice. According to the U.N. Climate Change Secretariat, all Parties must provide information on the following, as applicable to their NDCs:

- Quantifiable information on the reference point for GHG mitigation actions or targets;
- Time frame and/or periods (i.e., the start and end dates) for implementation;
- Scope and coverage of the NDC (i.e., the quantitative target, which sources and gases are covered);
- National planning processes for developing the NDC and, if available, implementation plans taking into account national circumstances;
- All assumptions and methodological approaches;
- How the Party determines that its NDC is fair and ambitious; and
- How the NDC contributes toward achieving the objective of the UNFCCC.[18]

The guidelines are to facilitate review and, under the committee (below), consultation intended to encourage compliance with commitments. Whether a Party supplies a timely NDC and reports its NDC according to the guidelines is subject to review by a technical group of experts. The adequacy and appropriateness of Parties' NDCs are not subject to review under the ETF.

---

[18] See U.N. Climate Change Secretariat, "FAQ's on the Operationalization of the Enhanced Transparency Framework," https://unfccc.int/enhanced-transparency-framework.

Flexibility in reporting under the PA is afforded only for those provisions in the modalities, procedures, and guidelines that are specified to allow flexibility. These provisions include (1) the frequency and level of details of reporting, (2) the modalities of the review, and (3) the modalities of the facilitative multilateral consideration of progress. A Party may determine whether to make use of flexibilities. That said, using the flexibilities is not without checks in the review processes.

A developing country that claims inadequate capacity to meet the guidelines and elects to apply a flexibility must make clear in its Biennial Transparency Report that it has applied a flexibility. It must explain the capacity constraint, how it intends to address the constraint, and its intended time frame to make improvements to the constraint(s).[19] The technical review teams may not review these flexibilities.

Committee

The Parties established a 12-member committee to "facilitate implementation" of the PA. The committee is intended to support Parties' efforts to meet their obligations under the PA as a soft, pro-compliance mechanism. The PA's compliance processes are consultative, not punitive. The committee may initiate a "consideration" should a Party not submit or update its NDC as required or provide mandatory communications.

Financing

In 2009 and 2010, developed countries pledged collectively to mobilize US$100 billion per year by 2020, from public and private sources, to support mitigation and adaptation activities in low- income countries. COP decision 1/CP.21 to adopt the PA (not the PA itself) stated that developed countries intend to continue their existing collective mobilization goal through 2025. Prior to 2025, the Parties shall set a new collective quantified goal for financial resources from a floor of US$100

[19] Decision 18/CMA.1, para. 6, 2018.

billion per year. The goal should take into account the needs and priorities of developing countries. Parties may take into consideration the information from the Warsaw International Mechanism for Loss and Damage associated with Climate Change Impacts.

The financial pledges are not an enforceable commitment by developed country Parties. Many stakeholders argue, nonetheless, that the resources are essential to help low-income countries contribute to GHG abatement and adaptation in the context of sustainable development. The financial flows are also important politically—in part to build confidence in the functionality of the UNFCCC and PA and to build trust between the lower and higher income economies.

At COP24 and since then, some countries made pledges toward this goal. Some developing country Parties submitted NDCs with GHG mitigation targets they would achieve unconditionally and more ambitious targets that they would achieve with adequate financial and technical support.

The Green Climate Fund (GCF) was proposed, during the 2009 COP in Copenhagen to be a new international financial institution connected to the UNFCCC.[20] The fund and its design was agreed during the 2011 COP in Durban, South Africa. The GCF was made operational in 2014. The GCF aims to assist lower-income countries in their efforts to combat climate change through the provision of grants and other concessional financing for mitigation and adaptation projects, programs, policies, and activities. The GCF is capitalized by contributions from donor countries and other sources, potentially including innovative mechanisms and the private sector.

The GCF officially opened for capitalization at the U.N. Climate Summit in September 2014. The GCF's initial resource mobilization lasted from 2015 to 2018.

---

[20] For more information on the GCF and other financial mechanisms under the UNFCCC, see CRS Report R41889, *International Climate Change Financing: The Green Climate Fund (GCF)*, by Richard K. Lattanzio and related products listed below.

As of the most recent published reporting (April 30, 2019), the GCF had raised over $10.2 billion in signed pledges from 48 countries/regions/cities during the resource mobilization period. The GCF board recently approved 10 new projects, increasing the GCF portfolio to 111 projects and increasing the level of related GCF funding to over $5.2 billion in 99 developing countries. On October 25, 2019, during the Pledging Conference for GCF's First Replenishment in Paris, 27 countries made pledges totaling $9.8 billion to cover the next four years of the fund.

Parties agreed in Katowice that the Adaptation Fund, which was established under the KP, will serve the PA, in addition to the Global Environmental Facility[21] and the GCF. Thus far, the Adaptation Fund has been financed by a share of the proceeds of the emissions trading mechanisms under the KP, as well as by voluntary contributions. With a transition from the KP to the PA, arrangements for the flow of funds are not completely agreed upon. Parties agreed that a share of the proceeds from one of the new cooperative mechanisms will continue to provide a share of its proceeds to the Adaptation Fund. A number of Parties oppose proposals—particularly from Parties that are relatively small and perceived to be especially vulnerable to climate change—to use the other two market mechanisms under Article 6 to finance the Adaptation Fund.

Beginning in 2020, developed countries are to submit biennial communications on expected levels of climate finance. The communications are to contain both quantitative and qualitative information. The biennial communications and Secretariat synthesis of the information therein is to inform the Global Stocktakes.

Starting in 2020, the Standing Committee on Finance is to report on the determination of support needs of developing countries to implement the UNFCCC and the PA. The committee is also to consider financial needs consistent with long-term low-emissions and sustainable development pathways.

---

[21] For more information, see CRS In Focus IF10144, *The Global Environment Facility (GEF)*, by Richard K. Lattanzio.

Technology

The Technology Framework of the PA is to provide overall guidance to the Technology Mechanism that was established under the UNFCCC. The purpose of both is to foster sharing of information and cooperation to develop new, low-emission technologies and technologies to increase resilience to climate change. Supporters viewed the technology mechanisms as important in transforming the set of technologies available, and the economies that use them, as a means to meet the objective of the UNFCCC.

The Technology Framework is to have five focus areas: (1) innovation, (2) implementation, (3) enabling environments and capacity-building, (4) collaboration and stakeholder engagement, and (5) support. The Parties intend that the framework should facilitate the active participation of all relevant stakeholders and take into account sustainable development, gender, the special circumstances of the least developed countries and small island developing states, and the enhancement of capacities of indigenous people and "endogenous technologies." The Executive Committee of the Technology Framework is expected to report on the progress and challenges of its work in joint annual reports with the Climate Technology Centre established under the UNFCCC.

In: Climate Change
Editor: Iyana Hendricks
ISBN: 978-1-53617-757-2

Chapter 8

GREEN CLIMATE FUND: RESOURCE MOBILIZATION, RECENT PROJECTS*

Richard K. Lattanzio

RESOURCE MOBILIZATION

The Green Climate Fund (GCF) was officially opened for capitalization at the U.N. Climate Summit in September 2014. (See CRS In Focus IF10382, *The Green Climate Fund (GCF)* for a history and overview of the fund.) The GCF's initial resource mobilization (IRM) lasted from 2015 to 2018. As of the most recent reporting, April 30, 2019, the GCF had raised over $10.2 billion (all amounts are in U.S. $ equivalent) in signed pledges from 48 countries/regions/cities during the IRM period. *Table 1* shows the pledges of the top 20 donor parties and the funds contributed as of April 30, 2019.

* This is an edited, reformatted and augmented version of Congressional Research Service Publication No. IF10762, Updated November 21, 2019.

Table 1. Top Pledges for GCF's IRM
(in millions U.S. $ equivalent, as of April 30, 2019)

Party	Pledged	Contributed
United States	3,000.0	1,000.0
Japan	1,500.0	1,500.0
United Kingdom	1,211.0	1,211.0
France	1,035.5	1,035.5
Germany	1,003.3	1,003.3
Sweden	581.2	581.2
Canada	277.0	277.0
Norway	272.2	272.2
Italy	267.5	267.5
Australia	187.3	187.3
Spain	160.5	56.2
Netherlands	133.8	133.8
Finland	107.0	76.1
Republic of Korea	100.0	100.0
Switzerland	100.0	100.0
Denmark	71.8	71.8
Belgium	66.9	66.9
Luxembourg	46.8	40.1
Austria	34.8	34.8
Mexico	10.0	10.0

Source: Green Climate Fund, https://www.greenclimate.fund/how- we-work/resource-mobilization.

Notes: United States dollars equivalent (U.S. $ eq.) based on the reference exchange rates established for GCF's High-Level PledgingConference (GCF/BM-2015/Inf.01).

First Replenishment

On October 25, 2019, during the Pledging Conference for GCF's First Replenishment in Paris, 27 countries made pledges totaling $9.8 billion for the next 4 years (see *Table 2*). No regions or cities made pledges on October 25, 2019.

Table 2. Pledges for GCF's First Replenishment (in millions U.S. $ equivalent, as of October 25, 2019)

Party	**Pledge**
United Kingdom	1,851.9
France	1,794.1
Germany	1,689.8
Japan	1,500.0
Sweden	852.6
Norway	433.7
Italy	350.2
Canada	229.2
Korea	200.0
Spain	176.5
Switzerland	155.5
Netherlands	141.1
Denmark	126.0
Finland	116.7
Belgium	46.7
Luxembourg	46.7
Austria	35.8
New Zealand	10.7
Ireland	4.8
Monaco	4.4
Poland	3.2
Slovak	2.3
Iceland	2.1
Slovenia	1.2
Portugal	1.2
Hungary	0.7
Indonesia	0.5
Liechtenstein	0.1

Source: Green Climate Fund.

ARREARS

Of the $10.2 billion signed pledges under the IRM, all funds have been deposited with the GCF as of April 30, 2019, except those shown in *Table 3*.

Table 3. Arrears for GCF's IRM (in millions U.S. $ equivalent, as of April 30, 2019)

Party	IRM Arrears
United States	2,000.0
Spain	104.3
Finland	31.2
Luxembourg	6.7
Malta	0.2
Indonesia	0.1

Source: Green Climate Fund.

RECENTLY APPROVED GCF PROJECTS

At its 23rd meeting (B.23), the GCF board approved 10 new projects and programs (see *Table 4*). In all, the total amount of GCF funding allocated for projects approved at B.23 was $266.9 million, which with planned cofinancing is to support over $1.4 billion in climate mitigation and adaptation in 10 developing countries. These new approvals increased the GCF portfolio to 111 projects and increased the level of related GCF funding to over $5.2 billion in 99 developing countries.

Table 4. List of New Projects Approved at B.23
(Project number, country, description, GCF funding)

FP107	*Kingdom of Bhutan:* The project aims to address the multifaceted challenges of climate change and enhance resilience of smallholder farmers in eight districts in Bhutan, through an integrated approach to climate resilient agriculture. \$25.3 million.
FP108	*Pakistan:* The objective of the project is to transform agriculture in the Indus Basin by increasing resilience among the most vulnerable farmers and strengthening the government's capacity to support the communities to adapt. \$35.0 million.
FP109	*Timor-Leste:* The project aims to safeguard vulnerable communities and their physical assets from climate-change-induced disasters by addressing institutional, financial, and legislative barriers and implementing rural infrastructure and catchment rehabilitation interventions to increase the climate resilience of the vulnerable rural population in six municipalities. \$22.4 million.
FP110	*Ecuador:* The project presents Ecuador's Reducing Emissions from Deforestation and Degradation results for 2014, where a total volume of 3.6 million metric tons of carbon dioxide equivalent emissions reductions are presented to the GCF for results-based payments. \$18.6 million.
FP111	*Honduras:* This project targets improvements in the provision of water services by increasing the resilience of vulnerable coniferous forests, which will be achieved by implementing forest restoration, agroforestry practices, and adaptive forest management in critical areas. \$35.0 million.
FP112	*The Marshall Islands:* The project supports the Republic of the Marshall Islands in adapting to increasing climate risks, particularly more frequent and extreme droughts, which impact the country's drinking water supply. \$18.6 million.
FP113	*Kenya:* The project aims to increase the resilience of the livestock sector and other land- use sectors in restored and effectively governed rangeland ecosystems in Kenya's arid and semi- arid lands. \$23.2 million.
FP114	*Ghana:* The program aims to empower vulnerable women groups in the country's most vulnerable agricultural zone, through a line of credit and technical assistance to participate in low-emission climate resilient agricultural practices. \$20.0 million.

Table 4. (Continued)

FP115	*Chile:* The Espejo de Tarapacá Project is comprised of two large-scale power plant projects that will provide clean renewable baseload energy supply: (1) a 300-megawatt pumped storage hydroelectric plant using the Pacific Ocean as its lower reservoir and an existing natural concavity as its upper reservoir, and (2) a 561-megawatt photovoltaic solar plant. $60.0 million.
SAP007	*Zimbabwe:* The project aims at directly supporting long-term adaptation of vulnerable and food insecure households (50,000 people, of which 66% are women) in Masvingo and Rushinga districts in Zimbabwe. $8.9 million.

Source: Green Climate Fund, "GCF, 23rd Meeting of the Board of the Green Climate Fund Summary of the Outcomes 6-8 July 2019, Songdo, Republic of Korea."

Notes: For a complete list of board-approved projects with full documentation and a list of accredited agencies, see the GCF website at http://www.greenclimate. fund/what-we-do/projects-programmes. All dollar amounts in U.S. $ equivalent.

INDEX

A

B

C

D

E

F

G

K

L

M

N

O

S

T

U

V

W